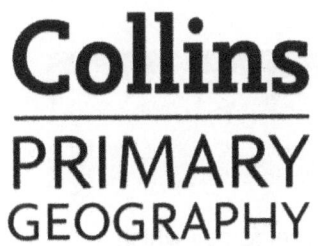

Change
Teacher's Guide 5

Stephen Scoffham | Colin Bridge

Geography in the primary school	2
Collins Primary Geography overview	3
Places, themes and skills	4
Layout of the units	6
Lesson planning	8
Lesson summary	9
Studying the local area	10
Studying places in the UK and wider world	11
Differentiation and progression	12
Assessment	13
Ofsted and the National Curriculum in England	14
Support and guidance	15
Unit-by-unit notes	16
Photocopiable resource matrix	26
Photocopiable resources	30
Geography in the National Curriculum in England	60
World maps	62

Geography in the primary school

Geography is the study of the Earth's surface. It helps children understand the human and physical forces which shape the environment and the way it is changing. Children are naturally interested in their immediate surroundings. They also want to know about places beyond their direct experience. Geography is uniquely placed to satisfy this curiosity.

Geographical enquiries

Geography is an enquiry-led subject that explores fundamental questions such as:

- Where is this place?
- What is this place like (and why)?
- How and why is it changing?
- How does this place compare with other places?
- How and why are places connected?

These questions involve not only finding out about the natural processes which have shaped our environment; they also involve finding out how people have responded to them. Studying this interaction at a range of scales from the local to the global and asking questions about what is happening in the world around us lies at the heart of both academic and school geography.

Geographical perspectives

Geographical perspectives offer a uniquely powerful way of seeing the world. Since the time of the Ancient Greeks, geographers have been attempting to chronicle and interpret their surroundings. One way of seeing links and connections is to think in terms of key concepts and ideas. Three concepts, which have proved particularly useful in a range of settings, are place, space and scale.

- Place focuses attention on specific places (real and imagined) and highlights their character, current activities, changes and development.
- Space focuses attention on the relationship between features and places, and refers to where they are located, the patterns they form and networks connecting them.
- Scale enables geographers to look at the world from very small local sites to international regions.

A layer of additional concepts provides a further way of enhancing geographical understanding. These concepts include pattern, change, movement, interconnections, culture, power, sustainability and environmental impact. Taken in combination, these concepts act as a 'lens' for describing and analysing the complexities of the world around us.

As they conduct their enquiries and investigations geographers make use of some subject-specific skills. Foremost among these are mapwork and the ability to represent spatial information. Geographers also champion the use of digital data which enables them to portray changes and explore different scenarios. The use of maps, charts, diagrams, tables, sketches and other cartographic techniques – all of which allow us to visualise and better understand data – come under the more general heading of 'graphicacy'. Graphicacy is sometimes seen as a key human attribute and a distinguishing feature of geographical thinking.

Geography in primary schools gives children from the earliest ages a fascinating window onto the world. It embraces major concerns such as climate change, migration and biodiversity loss. The challenge for educators is to find ways of providing experiences and selecting content that will help children develop an increasingly deep understanding of the world around them.

Collins Primary Geography overview

Collins Primary Geography is a complete programme for pupils in the primary school and can be used as a structure for teaching geography from ages 5–11 and beyond. At its core are six Pupil Books, each of which has a linked Workbook. This Teacher's Guide provides teaching notes and photocopiable resources for each lesson. Editable Word, PDF and PowerPoint files are available to help you adapt the resources to the needs of your class. Audio files are also available for the stories in Pupil Books 1 and 2.

Aims

The overall aim of the programme is to inspire children with an enthusiasm for geography and to empower them as learners. The underlying principles include a commitment to international understanding in a more equitable world; a concern for the future welfare of the planet; and a recognition that creativity, hope and optimism play a fundamental role in lasting learning. Three different dimensions – connecting to the environment, connecting to each other and connecting to ourselves – are explored throughout the programme in different contexts and at a range of scales. We believe that learning to think geographically in the broadest meaning of the term will help to prepare children for the future and whatever it may hold.

Structure

Collins Primary Geography provides full coverage of the National Curriculum in England framework. Each Pupil Book covers a balanced range of themes and topics and includes case studies with a more precise focus:

- Book 1 *World around me* introduces children to the world at a local scale.
- Book 2 *Our planet* explores the world at a global scale.
- Book 3 *Investigation* encourages pupils to conduct their own research and enquiries.
- Book 4 *Movement* considers how movement affects the physical and human environment.
- Book 5 *Change* includes case studies on how places alter and develop.
- Book 6 *Issues* considers more complex ideas to do with the environment and sustainability.

Although the books are not limited to a specific age group, Book 1 will be particularly suitable for children at the beginning of their formal education. Book 2 is suitable for ages 6–7, Book 3 for ages 7–8, Book 4 for ages 8–9, Book 5 for ages 9–10 and Book 6 for ages 10–11, or children at the end of primary school.

The programme is structured in such a way that key themes are revisited, making it possible to investigate a specific topic in greater depth if required.

Investigations

Enquiries and investigations are an important part of pupils' work in primary geography. Asking questions and searching for answers can help children develop core knowledge, understanding and skills. Fieldwork is time-consuming when it involves travelling to distant locations, but local area work can be equally effective. Many of the exercises in *Collins Primary Geography* focus on the classroom, school building and local environment. We believe that such activities can have a seminal role in promoting long-term positive attitudes towards sustainability and the environment.

Places, themes and skills

Collins Primary Geography Books 3 to 6 follow a structure that gives a balance between places, themes and skills. In the opening units, pupils are introduced to topics including the physical geography of Planet Earth, water, weather, settlements, work and travel, and the environment. The units that follow take a global perspective and highlight a range of case studies focusing on the UK, Europe, North or South America, and Africa or Asia. Key geographical skills such as mapwork and fieldwork are featured in all the units. The overall aim is to provide a balanced coverage of geography.

Places

Locality studies are featured in each unit. These studies illustrate how people interact with their physical surroundings in a constantly changing world. They draw on first-hand accounts, focus on contemporary issues and highlight successful classroom activities. The places have been carefully selected to enable pupils to develop a framework of reference points which will enable them to place new knowledge in context by the time they have completed the scheme.

Themes

Physical geography is covered in the initial three units of each book which focus on Planet Earth, water and weather. Human geography is considered in units on settlements, and work and travel. There is also a unit specifically devoted to the urban and rural environment and human impact on the natural world. This is a very important aspect of modern geography and a key topic for schools generally. Climate change is considered in all the units to draw attention to its effects in many areas of our lives. Pupil Book 6 includes a dedicated unit exploring climate change and what we can do about it. When teaching children about climate-change issues, it is important that children are not left feeling helpless. Learning about current issues is the first step towards constructive engagement and action.

Skills

Maps and plans are introduced in context to convey information about the places being studied. The books contain maps at scales which range from the local to global. Charts, diagrams and other graphical devices are included throughout to illustrate a variety of techniques which children can emulate. Fieldwork is strongly emphasised, and all the books include projects, investigations and mapwork exercises which can be conducted in the local environment. Please note the usual fieldwork health and safety considerations before undertaking these activities.

Cross-curricular links

The different units in *Collins Primary Geography* can be easily linked with other subjects. The physical geography units have natural synergies with themes from sciences, as do the units on the environment. Local area studies overlap with work in history. Furthermore, the opportunities for promoting the core subjects are particularly strong. For example, the interpretation and presentation of data in various tables, graphs and models has clear links to work in mathematics. Each lesson is centred around discussion questions, and many of the investigations involve written work in different modes and registers.

Oracy and critical thinking

Each lesson provides an opportunity for pupils to engage with and explore information and ideas through discussion. Discussion panels present questions that are graded, with opening questions tending to be factual and later questions requiring critical thinking or personal input. These activities give all pupils the opportunity to practise speaking with confidence, explaining their ideas, listening and responding to others, and participating in group discussions. You, as the teacher, can facilitate whole-class, group or one-on-one discussions by modelling speaking and asking prompting questions. There are also opportunities for pair and groupwork in some of the mapwork and investigation exercises, as suggested in the unit-by-unit notes. These will further encourage development of oracy and critical thinking.

Places and themes for Books 3 to 6

Places and themes	Book 3 units	Book 4 units	Book 5 units	Book 6 units
Planet Earth	Landscapes	Coasts	Seas and oceans	Restless Earth
Water	Water around us	Rivers	Wearing away the land	Drinking water
Weather	Weather worldwide	Weather patterns	The seasons	Climate change
Settlements	Villages	Towns	Cities	Planning issues
Work and travel	Travel	Food and shops	Jobs	Transport
Environment	Caring for nature	Caring for towns	Pollution	Conservation
United Kingdom	Scotland	Northern Ireland	Wales	England
Europe	France	Germany	Greece	Europe
North and South America	South America	North America	North America	South America
Asia and Africa	Asia	Asia	Africa	Asia

Layout of the units

Books 3 to 6 each have ten units divided into three lessons. In earlier units, pupils are introduced to key themes based around Planet Earth, water, weather, settlements, work and travel, and the environment at increasing levels of complexity. Later units focus on places from around the UK, Europe, North or South America, and Africa or Asia. The overall aim is to provide a balanced coverage of geography.

Each lesson then follows a consistent layout, with several recurring features, as follows:

Unit title

Identifies the focus of the unit across the three lessons.

Lesson title

Identifies the theme of the lesson. The supporting Workbook unit and photocopiable resource use the same title which makes them easy to identify.

Enquiry question

Presents a focusing question for a whole section of the lesson, and suggests opportunities for open-ended investigations and practical activities.

Key words panel

Highlights key geographical words and terms which will be used during the lesson. Introduce these words as you teach the lesson. Use discussion to reinforce understanding. Children could build up geography notebooks over the course of a year (or longer). These might include key terms with supporting illustrations and/or definitions, if appropriate. These will be a valuable record of pupils' work and development.

Introductory text

Provides a simple introduction to the lesson's focus, presenting key knowledge opportunities. (The unit-by-unit notes provide further topic information to give you, as the teacher, additional context for teaching each lesson. These notes are not designed for the pupils).

Discussion panel

Consists of questions designed to draw pupils into the topic and to stimulate discussion. The first question often involves simple comprehension. Other questions involve reasoning and/or introduce a human element which helps to relate the topic to the child's own experience. Guidance on encouraging oracy and critical thinking through the facilitation of high-quality discussion is given in the unit-by-unit notes.

Photographs and graphics

Photos and illustrations provide opportunities for exploration, demonstration and discussion. Graphical devices ranging from maps to satellite images amplify the topic.

Data bank

Provides extra information to engage children and encourage them to find out more for themselves. Pupils can research additional facts and figures for themselves, use them in a quiz or game, or simply add them to their geography notebook or a class display.

Climate change panel

Provides extra information and further discussion opportunities to show children the effects of climate change in many areas of our lives.

Mapwork exercise

Indicates how the lesson can be developed through atlas and mapwork.

Investigation panel

Suggests a practical activity which will help pupils consolidate their understanding.

Summary panel

Indicates the knowledge and understanding covered in the unit.

Photocopiable resources

Each lesson has a supporting photocopiable resource. These explore the concepts and ideas which underpin the unit, and extend what is presented in the book in engaging and informative ways. The resources may also be used to consolidate and assess understanding of the key concepts. See pages 30–59 of this book.

Workbook

The Workbook includes additional activities for each lesson, supporting learning, providing scaffolding for investigations, encouraging pupils to apply their learning and allowing teachers to provide evidence of learning in geography for each child.

Layout of the units

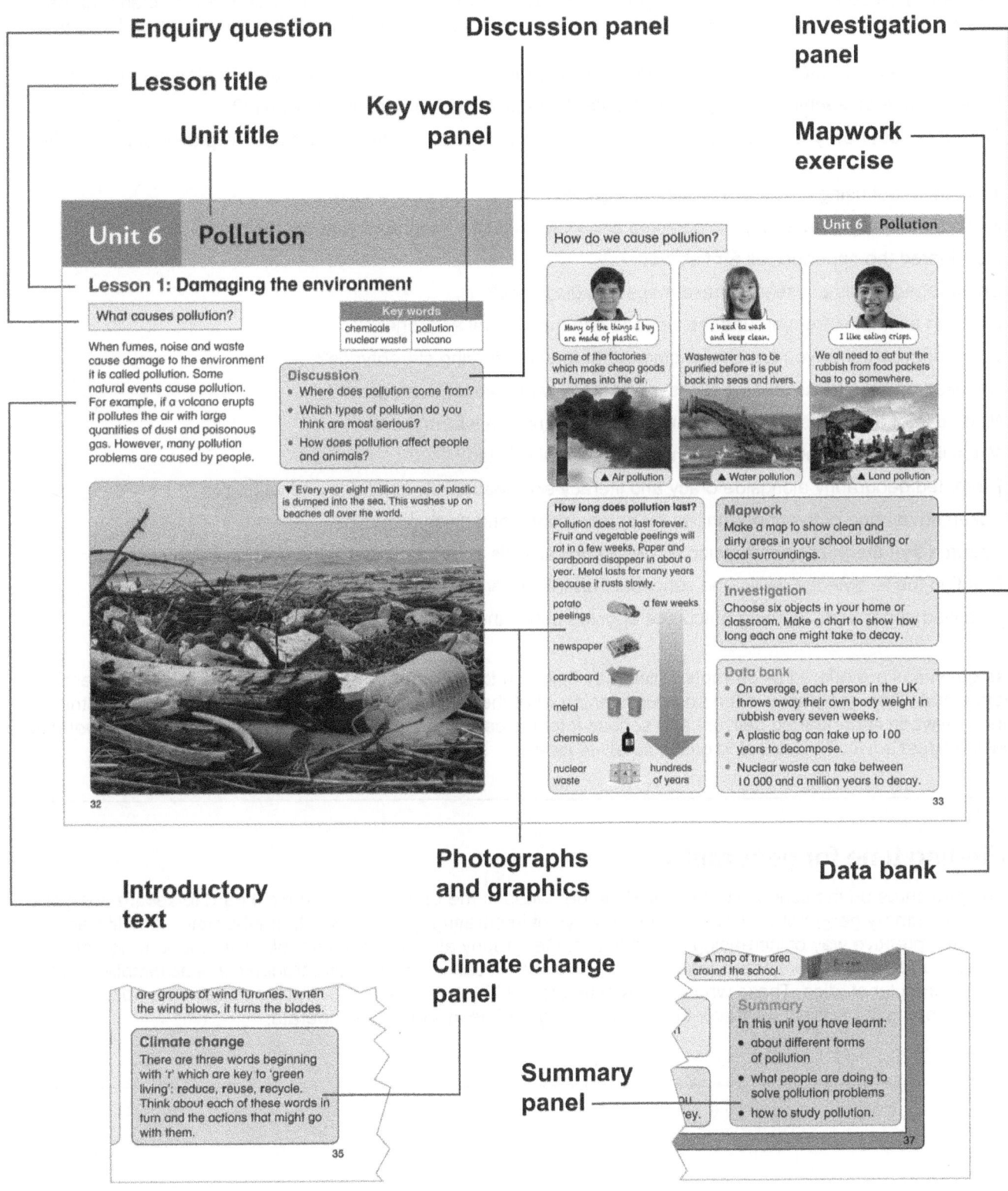

Lesson planning

Collins Primary Geography has been designed to support both whole-school and individual lesson planning. As you devise your schemes and work out lesson plans, you may find it helpful to ask the following questions. Have you:

- Given children a range of entry points which will engage their enthusiasm and capture their imagination?
- Used a range of teaching strategies which cater for pupils who learn in different ways?
- Thought about how you will draw pupils into class discussion, supporting them to develop their understanding and ideas?
- Thought about using practical activities and games?
- Explored the ways that stories or personal accounts might be integrated with the topic?
- Considered the opportunities for fieldwork?
- Encouraged pupils to use and make maps and diagrams?
- Included examples from around the world to enhance global awareness?
- Questioned whether you are challenging rather than reinforcing stereotypes?
- Checked on links to suitable websites, particularly with respect to research?
- Made use of digital programs to record findings or analyse information?
- Made links to other subjects where there is a natural overlap?
- Promoted geography alongside oracy and literacy skills especially in talking and writing?
- Taken advantage of the opportunities for presentations and class displays?
- Ensured that the pupils are developing geographical skills and meaningful subject knowledge?
- Clarified the knowledge, skills and concepts that will underpin the lesson?
- Identified appropriate learning outcomes or given pupils the opportunity to identify their own ones?

These questions are offered as prompts which may help you to generate stimulating and lively lessons. There is clear evidence that when geography is fun and pupils enjoy what they are doing, it can lead to lasting learning. Striking a balance between light-hearted delivery and serious intent is part of the craft of being a teacher. *Always remember to follow the latest advice for practising online safety in research activities.*

Finding time for geography

The pressures on the school timetable and the demands of the core subjects make it hard to secure adequate time for primary geography. However, finding ways of integrating geography with mathematics and literacy can be a creative way of increasing opportunities. Geography also has a natural place in a wide range of social studies and current affairs whether local or global. It can be developed through class assemblies and extra-curricular studies. Those who are committed to thinking geographically find a surprising number of ways of developing the subject whatever the accountability regime in which they operate.

Lesson summary

The table below provides an overview of the lessons in *Collins Primary Geography Pupil Book 5*. Individual schools may want to adapt the lessons and associated activities according to their particular needs and circumstances.

Theme	Unit focus	Lesson 1	Lesson 2	Lesson 3
Planet Earth	Seas and oceans	Beneath the surface	The ocean environment	Learning about seas
Water	Wearing away the land	Rivers in action	Preventing flood damage	Finding out about rivers
Weather	The seasons	Changing seasons	Seasons worldwide	Seasonal influences
Settlements	Cities	Describing cities	World cities	The story of London
Work and Travel	Jobs	Making things	Different jobs	Types of work
Environment	Pollution	Damaging the environment	'Green living'	Exploring clean energy
United Kingdom	Wales	Mountains and valleys	The story of Blaenavon	A visit to Big Pit
Europe	Greece	Introducing Greece	Summer in Athens	A Greek island
North and South America	North America	Introducing the Caribbean	Finding out about Jamaica	Living in Jamaica
Asia and Africa	Africa	Introducing Africa	Learning about Kenya	Living in Kenya

Studying the local area

The local area is the immediate vicinity around the school and the home. It consists of three different components: the school building, the school grounds, and local streets and buildings. By studying their local area, children will learn about the different features which make their environment distinctive and how it attains a specific character. When they are familiar with their own area, they will then be able to make meaningful comparisons with more distant places.

There are many opportunities to support the lessons outlined in *Collins Primary Geography* with practical local area work. First-hand experience is fundamental to good practice in geography teaching, is a clear requirement in the programme of study and has been highlighted in guidance to Ofsted inspectors. The local area can be used not only to develop ideas from human geography but also to illustrate physical and environmental themes. The checklist below illustrates some of the features which could be identified and studied.

Physical geography

Hill, valley, cliff, mountain, rock, slope, soil, forest

River, stream, pond, lake, estuary, ocean, sea, beach, coast

Slopes, rock, soil, vegetation and other small-scale features

Local weather, seasons and site conditions

Human geography

Origins of settlements (city, town, village), land use (farms) and economic activity

House, cottage, terrace, flat, housing estate

Roads, stations, harbours, ports

Shops, factories and offices

Fire, police, ambulance, health services

Library, museum, park, leisure centre

All work in the local area involves collecting and analysing information. An important way in which this can be achieved is through the use of maps and plans. Other techniques include annotated drawings, bar charts, tables and reports. There will also be opportunities for the children to make presentations in class and perhaps to the rest of the school in assemblies.

Misconceptions

There is a growing body of research which helps practitioners to understand more about how children learn primary geography and the barriers and challenges that they commonly encounter. The way that young children assume that the physical environment was created by people was first highlighted by Jean Piaget. The importance and significance of early childhood misconceptions was further illuminated by Howard Gardner. More recent research has considered how children develop their understanding of maps and places. Children's ideas about other countries and their attitudes to other nationalities form another very important line of enquiry. So, too, do their ideas about climate change and the environment. Some key readings are listed in the references on page 15.

Studying places in the UK and wider world

Collins Primary Geography Pupil Book 5 contains detailed case studies of the following places in the UK and around the world. Place studies focus on small-scale environments and everyday life, which means they relate to children's needs and understanding.

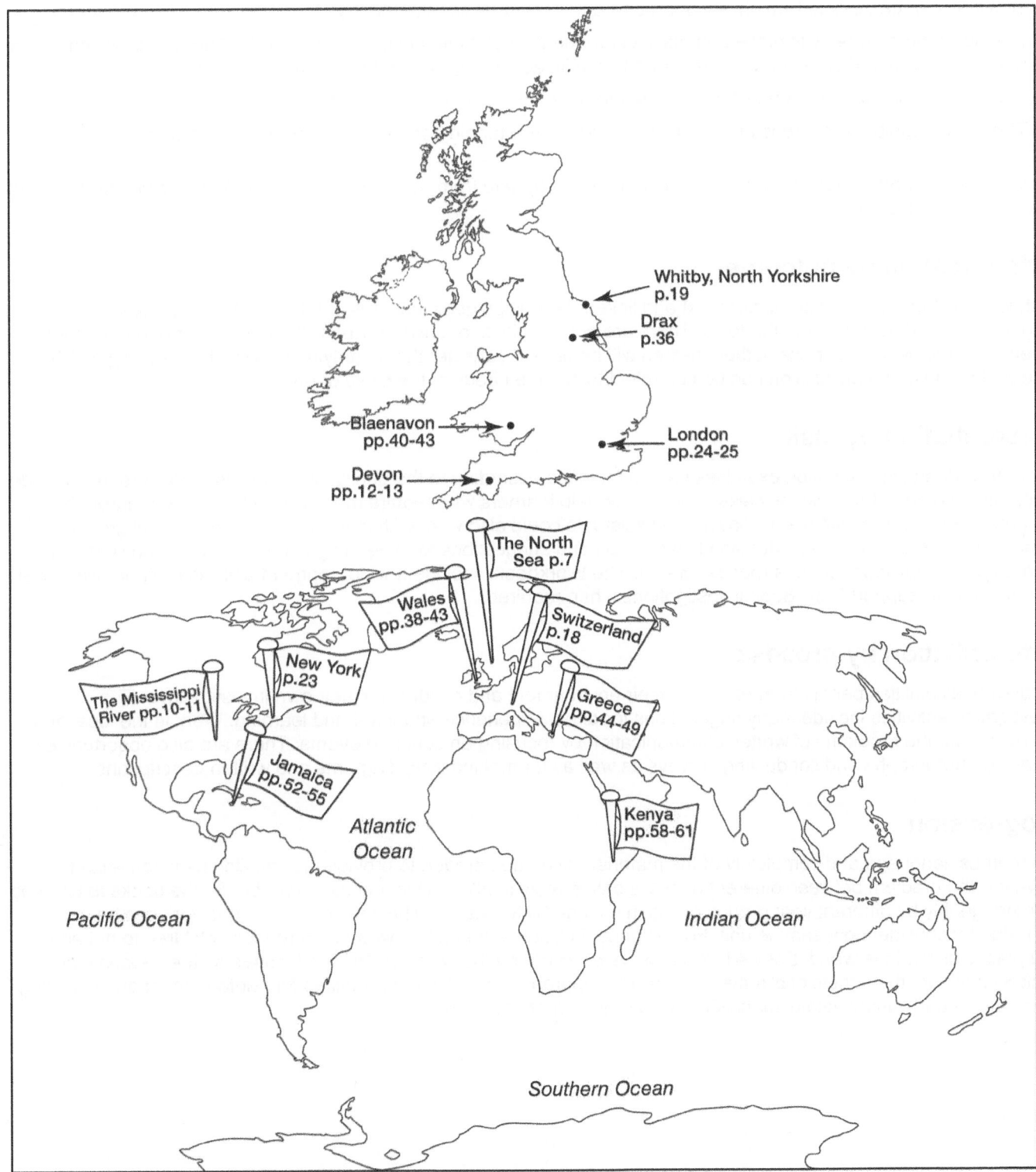

Further examples from around the world are also included throughout the units. By considering people and describing their surroundings, the information is presented at a scale and in a manner which relates particularly well to children. Research shows that pupils tend to reach a peak of friendliness towards other countries and nations at about the age of 10. It is important to capitalise on this educationally and to challenge prejudices and stereotypes.

Differentiation and progression

Collins Primary Geography sets out to provide access to the curriculum for children of all abilities. It is structured so that children can respond to and use the material in a variety of ways.

- Each lesson contains a range of stimulus material designed to engage children imaginatively. This means activities can be selected which are appropriate to individual circumstances.
- Considerable emphasis is placed on discussion and critical thinking appropriate for the age group, which allows teachers to frame discussions and respond to pupils according to their level of understanding.
- The Workbooks provide further opportunities to develop and evidence understanding.
- The photocopiable resources allow you to assess understanding of key concepts during the unit.

Teachers will be able to select what they think will be appropriate from a range of resources. There is no need to work through all the material.

Differentiation by outcome

Each lesson starts with an introductory text and linked discussion questions designed to capture the children's imagination and draw them into the topic. There are opportunities for children who require more support to relate the material to their own experience. Other children will be able to consider the underlying geographical concepts. The pace and range of the discussion can be controlled to suit the needs of the class or group.

Differentiation by task

The mapwork and investigation exercises can be modified according to the pupils' ability levels. Teachers may decide to complete some of the tasks as class exercises or help learners who require more support by working through the first part of an exercise with them. Classroom assistants could also work with individual children or small groups. Some children could be given extension tasks. Ideas and suggestions for extending each lesson are provided in the information on individual lessons (pages 16–25). The Workbooks include a wider range of activities with varying levels of scaffolding to support Pupil Book investigations when required.

Differentiation by process

Children of all abilities benefit from exploring their environment and conducting their own investigations. The investigation activities include many suggestions for direct experience and first-hand learning. Work in the local area can overcome the problems of written communication by focusing on concrete events. There are also opportunities for taking photographs and conducting surveys as well as for making lists, diagrams and written descriptions.

Progression

The themes, language and complexity of the material have been graded to provide progression between each title. However, the gradient between different books is deliberately shallow. This makes it possible for the books to be used interchangeably by different year groups or within mixed-ability classes. The way that this might work can be illustrated by considering a sample unit. For instance, in Book 3 the unit on weather introduces children to hot and cold places around the world. Book 4 looks at ways of recording the weather; Book 5 focuses on the seasons, and Book 6 considers the effects of climate change. This approach provides opportunities for reinforcement and revisiting which will be particularly helpful for children who require more support.

Assessment

Assessment is often seen as having two different dimensions.

- *Formative assessment* is an ongoing process which provides both pupils and teachers with information about the progress they are making in a piece of work.
- *Summative assessment* occurs at defined points in a child's learning and seeks to establish what they have learnt and how they are performing in relation both to their peers and to nationally agreed standards.

Collins Primary Geography provides opportunities for both formative and summative assessment.

Formative assessment

- The discussion questions invite pupils to discuss a topic, relate it to their previous experience and consider any issues which may arise, thereby yielding information about their current knowledge and understanding.
- The mapwork exercises focus especially on developing spatial awareness and skills and will indicate the pupils' current level of ability.
- The investigation activities give pupils the chance to extend their knowledge in ways that match their current abilities.
- The Workbook activities provide an opportunity for pupils to apply their understanding and evidence learning.

Summative assessment

- The Summary panels at the end of each unit in the Pupil Book highlight key learning outcomes. These can be tested directly through individually designed exercises. These unit aims are repeated in the unit-by-unit notes in this Teacher's Guide.
- The photocopiable resources (see pages 30–59) can be used to provide further evidence of key concepts, and to track progress of knowledge and skills. Whether used formatively or summatively, they are intended to consolidate understanding and identify gaps in learning to inform future teaching.

Reporting to parents and guardians

Collins Primary Geography is structured around geographical skills, themes and places. As children work through the lessons, they can build up a folder of work and progress through the Workbook. This will provide evidence of mapwork and other practical activities both inside and outside the classroom and provide a rounded portrait of pupil achievement. This will also be a useful resource when teachers report to parents and guardians and show if a child requires further support in geography.

National curriculum reporting

There is a single attainment target for geography and other National Curriculum subjects in the National Curriculum for England. This simply states that

> 'By the end of each key stage, pupils are expected to know, apply and understand the matters, skills and processes specified in the relevant programme of study.'

This means that assessment need not be an onerous burden, and that evidence of pupils' achievement can be built up over a set of years (or a Key Stage). The assessment process can also inform lesson planning.

Ofsted and the National Curriculum in England

The National Curriculum in England provides a framework for geography but doesn't specify the details of what should be taught or in what depth. Schools have the flexibility to choose their own curriculum approaches provided they pay sufficient attention to (a) context (b) structure (c) sequencing and (d) implementation. There is a significant emphasis on factual knowledge. Ofsted argue that it is essential to identify what children need to remember and to use it transferably in different circumstances.

The curriculum

The curriculum refers to what is taught and should support children to build their knowledge over time. Ofsted distinguish between different forms of knowledge:

- *Substantive knowledge* refers to knowledge relating to the themes and topics specified in the National Curriculum. This could be seen as the 'vocabulary' of geography.
- *Disciplinary knowledge* involves applying a geographical 'lens' to a particular or area of study and can be regarded as the 'grammar' of geography.

If pupils haven't grasped substantive geographical knowledge, then they will be unable to think or speak geographically.

Lesson planning

Ofsted argues that pupils get better in geography by building on their prior knowledge and applying it in new more complex ways (Ofsted 2021). Activities should therefore be selected which help pupils to build their knowledge and consolidate what they have already learnt in time-efficient ways. This draws attention to the importance of sequencing in geography. It is important to think carefully not only about the building blocks of geography but also about what comes after as well as what comes before any particular topic. In this way the curriculum becomes the assessment model.

Inspection findings

Inspection findings indicate that practice is not always as good as it could be. Areas of weakness include mapwork, fieldwork, sequencing and the application of geographical concepts (Ofsted 2021, 2023). To some extent this is unsurprising given that limited time has been spent learning how to teach geography during primary training (Ofsted 2023). However, there is also clear evidence that pupils enjoy geography and are curious about the world around them. The fact many are passionate about the Earth and the need to care for it also attracted very favourable comment from inspectors (Freeland 2021).

These prompts may help you prepare for an inspection:

- Identify a teacher who is responsible for developing the geography curriculum.
- Decide how geography will fit into your whole school plan.
- Make an audit of current geography teaching to identify gaps and weaknesses.
- Discuss and develop a geography policy which includes statements on overall aims, topic planning, teaching methods, progression, assessment and recording.
- See that all members of staff are familiar with the geography curriculum.
- Organise in-service training to rectify any areas of weakness.
- Review and update geography teaching resources.
- Devise an action plan for geography which includes an annual review procedure.
- Discuss the policy with the school governors.
- Provide a regular opportunity for discussing geography teaching in staff meetings.

Support and guidance

Primary Geography Quality Mark

The Primary Geography Quality Mark set up by the UK Geographical Association is a self-assessment framework designed to help subject leaders. There are three categories of award. The 'bronze' level recognises that lively and enjoyable geography is happening in your school; the 'silver' level recognises excellence across the school; and the 'gold' level recognises that excellence that is shared and embedded in the community beyond the school. The framework is divided into four separate cells: (a) pupil progress and achievement; (b) quality of teaching; (c) behaviour and relationships; (d) leadership and management. For further details see the Geographical Association website.

Achieving accreditation for geography in school is a useful way of badging achievements and identifying targets for future improvement. This makes it an effective and efficient way of raising standards. The Geographical Association provides a wide range of support to teachers to help with this process. In addition to conferences and CPD sessions it produces a journal for primary schools, *Primary Geography*, three times a year. See the Geographical Association website for full details of the books and guides it publishes for classroom use.

Networking, training and sharing ideas

Networking and sharing ideas can happen on an informal basis amongst friends and professional colleagues. Conferences and CPD training and events provide a more formal way of developing and extending your knowledge of geography teaching. The support that comes from networking will help you to grow in confidence and broaden your ideas. Speaking or writing about what you have been doing will further consolidate your ideas. Subject associations, environmental organisations and development education centres usually welcome new members with enthusiasm (for example, see the Geographical Association and the Royal Geographical Society websites). The sense of community that they foster cannot be underestimated.

References and reading

Barlow, Anthony, and Sarah Whitehouse (2019) *Mastering Primary Geography*, London: Bloomsbury Academic.

Bonnett, Alistair (2023) *What is Geography?* (2nd edn), Lanham, Maryland: Rowman and Littlefield.

Cannell, Jon (2023) 'Geographical concepts in primary education', *Primary Geography*, 112: pp8–9.

Catling, S. et al. (2022) 'Aspiring to High-Quality Primary Geography: A report on a study of the GA's Primary Geography Quality Mark Moderators' feedback to schools', Sheffield: Geographical Association.

Dolan, Anne M. (2020) *Powerful Primary Geography: A Toolkit for 21st-Century Learning*, London: Routledge.

Freeland, Iain (2021) 'Geography in outstanding primary schools', Ofsted: schools and further education & skills (FES) blog, gov.uk.

Ofsted (2021) 'Research review series: Geography', gov.uk.

Ofsted (2023) 'Getting our bearings: geography subject report', gov.uk.

Roberts, Margaret (2023) *Geography Through Enquiry: Approaches to teaching and learning in the secondary school* (2nd edn), Sheffield: Geographical Association (Chapter 4).

Scoffham, Stephen (ed.) (2016) *Teaching Geography Creatively (Learning to Teach in the Primary School Series)* (2nd edn), London: Routledge.

Scoffham, Stephen and Paula Owens (2024) *Bloomsbury Curriculum Basics: Teaching Primary Geography* (2nd edn), London: Bloomsbury.

Tanner, Julia and Stephen Pickering (eds) (2017) 'Taking the Learning Outdoors at KS1', *Teaching Outdoors Creatively*, London: Routledge.

Trait, Georgie, et al. (2024) 'The key ingredients for quality geography', *Primary Geography*, 114: p19.

Willy, Tessa (ed.) (2019) *Leading Primary Geography: The essential handbook for all teachers*, Sheffield: Geographical Association.

Unit-by-unit notes

Unit 1: Seas and oceans

In this unit, pupils learn:
- about what lies beneath the surface of the ocean
- how oceans are threatened
- about conserving seas and oceans.

We tend to think of oceans as barriers because we live on dry land. Their vast size and depth make them seem difficult to penetrate. The average depth is over 3 kilometres, which means that the oceans go down much deeper than the land rises up. In places, there are dramatic landscape features, such as underwater mountains, ridges and valleys. We now know that the oceans play a major part in the global ecosystem. They stabilise the climate, balance the atmosphere and support a great variety of life. Recently, scientists have discovered new forms of life which feed on the sulfur and minerals that bubble out of volcanic vents. However, it is also established that the oceans are becoming more acidic as they absorb atmospheric pollution. This means they are masking the full impact of global warming.

Lesson 1: BENEATH THE SURFACE

What is it like under the oceans?
Most children will have no direct experience of the different plants and animals which live below the ocean surface. The drawing and photographs are designed to capture pupils' imagination and encourage discussion. They also show the changes which occur with increasing depth.

Discussion
Talk through the different levels and how each environment would affect the creatures found there. Below a few hundred metres it is almost completely dark, the water is just above freezing and the pressure begins to increase considerably.

Mapwork
The Great Barrier Reef was as a case study in Book 3. Find out about the distribution of coral reefs worldwide to highlight how they are found especially in the Caribbean, Indonesia, Philippines and Pacific Islands.

Investigation
You might structure the class scrapbook around sea creatures in general or a specific ocean, or get the children to select one image for each letter of the alphabet. Always follow the latest advice for practising online safety in research activities.

Lesson 2: THE OCEAN ENVIRONMENT

What are the threats to the ocean environment?
The photographs and text panels describe ways in which people exploit the oceans and also draw attention to environmental and conservation issues. Managing the ocean environment is complicated as it requires international agreement. Enforcing laws and agreements is another issue.

Discussion
Pupils can use discussion to build understanding in this activity. Their ideas will feed into the investigation. Make sure pupils are aware of what can be done to improve the ocean, for example using renewable energy, returning smaller fish and agreeing fishing limits.

Mapwork
You could provide pupils with a polar projection which shows the North Pole at the centre of a circular map or help them to find one in an atlas.

Investigation
Encourage pupils to use maps and charts as well as drawings in their posters.

Lesson 3: LEARNING ABOUT SEAS

What is a sea?

Discussion
Many people use the term 'sea' and 'ocean' interchangeably. However, seas are much smaller and shallower than oceans, and they are often partially enclosed.

Mapwork
For the exercise on Pupil Book page 6, remind pupils to include bays and gulfs as they make their list of seas.

The North Sea
The case study of the North Sea is effective because it fringes the east coast of the UK and also because of the enormously different ways in which it is used.

Mapwork
To support the activity on page 7, you could provide pupils with up-to-date maps of North Sea wind farms, which can be found online.

Teacher's Guide photocopiable resources
Use pages 30–32 to consolidate key concepts.

Workbook
See pages 2–7 for additional supportive activities.

Unit-by-unit notes

Unit 2: Wearing away the land

In this unit, pupils learn:
- that rivers are a major influence on the landscape
- how people try to control rivers
- how to study a river.

Landscapes slowly take shape over very long periods of time as a result of weathering and erosion. Weathering is the breakdown of rocks and other materials in the place where they are found. Erosion is the removal of these materials by wind, water, gravity or ice. Rivers play an important part in the process of erosion. The sheer force of the water as it flows downhill wears away the land. However, as rivers gather particles of rock, the water becomes much more abrasive and scrapes away at the sides and bottom of the channel. Further downstream, rivers deposit their load as mud and sediment.

Around the world, people exploit rivers in different ways. Dams are particularly helpful in preventing floods and providing water for drinking, drainage, generating power and irrigating fields. However, controlling rivers is a complicated business which entails environmental and financial costs as well as benefits.

Lesson 1: RIVERS IN ACTION

How do rivers shape the land?
The photograph of the canoeist negotiating a rapid illustrates the force of the water as it cascades downhill. Rapids are usually caused by layers of hard rock which wear away more slowly than the surrounding areas of softer rock. This gives the river bottom a continuous slope, unlike waterfalls which have a vertical drop.

Discussion
Children often find it difficult to understand the idea of erosion, transportation and deposition. These are complex processes that happen very gradually. However, they can also be seen much more vividly in times of flood when large objects, such as trees and boulders, are carried downstream. Start by explaining each process. Then ask children to explain them to the group to check understanding.

Investigation
Many of the terms which relate to rivers such 'source', 'channel' and 'mouth' are homonyms – words with more than one meaning. You might want to explain this to pupils so that they do not become confused.

Lesson 2: PREVENTING FLOOD DAMAGE

How can we control rivers?
The Mississippi was chosen for this case study not only because it is a major world river, but because it also causes spectacular floods. Until recently many people believed that it was possible to control the Mississippi, but now they are not so sure. This reflects a change of attitude towards the environment. Rather than seeking to dominate natural forces, environmentalists argue that we should work in harmony with them.

Climate change
Natural solutions to flooding aim to reduce the flow of water through the landscape. This can be done by planting trees and hedges and improving soil cover. In some places beavers have been re-introduced. This not only enhances wildlife; the dams that they build hold back the water when there is heavy rain. You may want to introduce this information to the children and begin a class discussion about ways to work with nature.

Lesson 3: FINDING OUT ABOUT RIVERS

What data is needed to find out about a river?
The work done by Avenue Primary School shows the educational potential of a river study. Fieldwork is important in geography, which many people argue is best learnt 'through the soles of your feet'.

Discussion
Even if you are unable to leave the classroom, the children can learn about rivers at second hand from the account given here. If you do decide to conduct a fieldwork study, you could use the eight questions in the survey on page 12 to structure the work. From a safety point of view, remember to include the risk of heavy rain and rising water levels in your risk assessment.

Mapwork
If you teach in the UK, this is a good opportunity to use a 1:50 000 Ordnance Survey map of your locality. Other maps will work depending on your location.

Investigation
Pupils can work in pairs to investigate and solve this problem through talk. Each of the survey questions could be developed as a more substantial study if you have time.

Teacher's Guide photocopiable resources
Use pages 33–35 to consolidate key concepts.

Workbook
See pages 8–13 for additional supportive activities.

Unit-by-unit notes

Unit 3: The seasons

In this unit, pupils learn:
- how the seasons are different
- about seasons around the world
- how people are affected by the seasons.

The seasons are caused by the movement of the Earth around the sun and the tilt in the Earth's axis. In June, the North Pole is tilted towards the Sun creating summer in the northern hemisphere and winter in the southern hemisphere. In December, the North Pole is tilted away from the sun and the seasons are reversed. In very general terms, the world can be divided into three main climate regions. There are cold regions around the North and South Poles, hot regions around the equator and Tropics, and temperate areas in between.

The UK lies in the temperate area and has four distinct seasons. These have a dominant effect on plant and animal life. Even in the modern world, the seasons also influence the crops people can grow, the houses we live in and the clothes we wear.

Lesson 1: CHANGING SEASONS

What are the seasons?
There are two key points in this lesson. First, pupils need to understand that the seasons follow a pattern. Second, they need to appreciate that whilst there may be slight variations in individual years, over a period of time the overall character of each season is clearly identifiable.

Discussion
The charts, diagrams and photographs provide different but complementary information about seasonal change and its impact on natural life. Check that children are able to 'read' the two charts by asking questions about temperature and sunshine from the data. Check they see the relationship between the frog's life cycle and the seasons. Then pupils can discuss the questions.

Investigation
Pupils could make individual seasons dials. You could also construct a large one as a class display.

Lesson 2: SEASONS WORLDWIDE

Do all places have the same seasons?
Children often believe that all places have the same pattern of seasons as they experience. The case studies challenge this assumption. Southeast Asia offers a striking contrast to the Mediterranean – here, a quarter of the world's population depends on the monsoon rains for their survival.

Discussion
Check pupils' advancing understanding as they discuss.

Mapwork
Monsoon climates are sometimes grouped under the more general heading of 'tropical climate'. The key feature is distinctive wet and dry seasons.

Investigation
The discussion work should form a bridge to this short writing game. Check that pupils are clear on the seasonal pattern in their country.

Lesson 3: SEASONAL INFLUENCES

How are farmers affected by the seasons?
All over the world, the work that farmers do is influenced by the seasons. There is a pattern to planting, growing and harvesting crops which is part of the culture in many countries. This lesson focuses on how farmers in the mountains of Switzerland keep cows and make milk. This involves 'transhumance', in which people and animals go to high pastures in the summer months and come back down to lower levels in the winter.

Climate change
Farmers are adapting to climate change by growing new crops and changing how they use the land. Discuss how this brings both threats and opportunities as in the Swiss cheese farmers example.

Mapwork
Depending on your location, you might ask pupils to plan a walk for a contrasting season too.

Investigation
Children might complete the seasons chart as a homework exercise, using their home as the example.

How are seaside resorts affected by the seasons?
Seasonal changes have a very significant impact on shops and businesses in seaside resorts. Whitby in Yorkshire is representative of places all around the coast of the UK in this respect.

Discussion
Encourage pupils to give examples and reasons as they discuss seasons, how Whitby is affected, and whether the current weather is typical.

Teacher's Guide photocopiable resources
Use pages 36–38 to consolidate key concepts.

Workbook
See pages 14–19 for additional supportive activities.

Unit-by-unit notes

Unit 4: Cities

In this unit, pupils learn:
- how cities are different from other places
- how cities are changing
- how to describe a city.

The first cities were built thousands of years ago in the Middle East, Egypt and China. They depended on a settled system of agriculture which produced enough food to support an urban population. As people congregated together, so art, politics, science and culture began to flourish. Today, cities dominate human affairs. For the first time in history, more than half the world's population now live in built-up areas.

Urban life brings many challenges. In many places, cities tend to be ringed by sprawling shanty towns. And in many countries, the decay of inner-city areas is a cause for concern, which redevelopment schemes have only partially addressed.

Lesson 1: DESCRIBING CITIES

What are cities like?
Children sometimes find it hard to distinguish between a city and a large town. Size is the key difference. Cities are large enough to have many different areas including business districts, suburbs, industrial areas and underground railways and airports.

Climate change
Discuss with the children some of the changes to combat climate change they would like to see in the place where they live.

Investigation
Try to get children to include both positive and negative aspects of city life as they complete this exercise.

Discussion
You could use this as a class discussion to draw together some of the pupils' ideas from the Investigation on cities, what it is like to live in them and why they are growing.

Lesson 2: WORLD CITIES

How are cities changing?
The world map shows that most of the world's largest cities are now the tropics or sub-tropics rather than in the mid latitudes. This is a trend which is likely to continue. So too is urban growth. It is estimated that three-quarters of the world's population could be living in cities by 2050.

Discussion
You could extend the discussion by considering the pattern which the world map reveals. How many cities are shown in the Northern as opposed to the Southern Hemisphere? Which continents have the most cities?

Investigation
This activity is an excellent opportunity to apply maths skills in context, using real data.

The story of New York
New York, along with London and Paris, is one of a dozen or so well-established world cities that had a population of over a million people in 1900.

Mapwork
Pupils could work together in groups with atlases, developing their group work and communication skills.

Lesson 3: THE STORY OF LONDON

How has London grown and changed?
Cities develop from smaller settlements and flourish for a variety of reasons. London has benefited from its proximity to the European mainland. However, it is rather poorly placed as a capital for the UK. Rather than being in the geographical centre of the country, it is around 100 miles to the southeast.

Investigation
Always follow the latest advice for practising online safety in research activities.

Mapwork
Pupils might take a theme such as museums or devise cycle or pedestrian routes when they draw their city route maps.

Climate change
In addition to London, Adelaide in Australia is the world's second National Park City. There are also settlements in many different countries which have been directly planned as garden cities. Talk with the children about what makes them special. Are there any in your area?

Teacher's Guide photocopiable resources
Use pages 39–41 to consolidate key concepts.

Workbook
See pages 20–25 for additional supportive activities.

Unit-by-unit notes

Unit 5: Jobs

In this unit, pupils learn:
- what factories do
- how jobs are linked together
- about how jobs can be grouped.

Factories make goods from natural resources or raw materials. They also assemble components or parts which have been made elsewhere. Usually, we only see the finished product and are largely unaware of the process by which goods are manufactured or the jobs involved in making them or their environmental impact. This unit introduces pupils to different types of work. One of the striking features of modern economies is that more and more people are involved in providing services. This means that it is increasingly hard to see how individual activities contribute to creating wealth.

Lesson 1: MAKING THINGS

Where are things made?
When you look at a factory from the outside, there is little way of knowing what happens inside. This lesson explores the way that factories are organised and identifies some of the different work areas that are needed.

Discussion
See if children can identify anything in the classroom made in factories. Check they understand that factories take raw materials and turn them into something to sell. You could maybe have a bit of fun with the idea of mass production in factories and schools – you could even have an input–output approach.

How do factories work?
The diagrams and text summarise the process of turning raw materials into products.

Investigation
The notion of inputs and outputs provides a generic way of thinking, which can be applied not only to factories, but to a range of other processes. Make links to other subjects, such as in mathematical puzzles.

Lesson 2: DIFFERENT JOBS

How do people earn a living?
Many children have rather confused notions about what work actually entails. This study of a small harbour illustrates how a range of activities combine to running and operating a workplace. The special skills which each person contributes are highlighted.

Discussion
Ask pupils to work in pairs to look closely at the picture to work out all the things that are going on. Point children towards the shipping containers stacked on the left that contain goods to be transported, the fishing boats with their catch, the ship delivering gravel with the crane and the lorry that will distribute it, and the workers who keep records of what goes in and out of the harbour. This will allow pupils to work out what a harbour does.

Mapwork
Pupils will have to imagine the overhead view in order to make the plan. This is an important mapwork skill.

Investigation
You could extend the activity by getting children to think about the technology that helps each person to perform their tasks.

Lesson 3: TYPES OF WORK

What are the different types of work?
Two hundred years ago, primary activities provided the majority of employment for people both in the UK and around the world. Most people worked on the land or supported farmers either directly or indirectly. As the economy has changed, so has the nature of work. Most jobs are now found in towns and cities. This has had profound effects on individual lifestyles, on where people live and the distance that they travel.

Discussion
Check that pupils understand the three categories by asking around the class for examples. You might then use pair work to discuss jobs at school. Ask pupils to challenge each other to explain and give examples.

Investigation
When the children compile their scrapbooks, they might make up some jobs of their own to add to the examples.

Mapwork
You can develop the activity by considering primary, secondary and tertiary activities in and around your school.

Teacher's Guide photocopiable resources
Use pages 42–44 to consolidate key concepts.

Workbook
See pages 26–31 for additional supportive activities.

Unit-by-unit notes

Unit 6: Pollution

In this unit, pupils learn:
- about different forms of pollution
- what people are doing to solve pollution problems
- how to study pollution.

There is nothing new about pollution. The streets of medieval London and other cities were littered with rubbish, and the royal court moved from place to place as sewage began to create a health hazard. What has changed, however, is the scale of the problem. In the last 50 years, industrial production has increased tenfold and human numbers have more than doubled. This has put the environment under immense strain. People are beginning to realise that the Earth is an enclosed ecosystem with a finite capacity but have not yet changed their lifestyles in response.

Lesson 1: DAMAGING THE ENVIRONMENT

What causes pollution?

Discussion
Point to the photographs and elicit the different types of pollution: air, water and land. Ask pupils to discuss and evaluate the photos, text and diagrams. They can build their understanding and debate. Extend by introducing plastic pollution. (Unlike air pollution, which is largely invisible, plastic pollution is apparent. Many beaches around the world are polluted. Plastic accounts for three quarters of beach litter. Plastic causes health problems and has a terrible effect on wildlife.) Children are often aware that plastic causes problems. Consider practical activities they can do, such as litter picks and surveys.

How do we cause pollution?
The three children illustrate that we are all responsible for pollution in different ways. The point is not to make children feel guilty but to realise there are often good reasons why we cause pollution in the first place.

Mapwork
The mapwork exercise is a reminder that pollution affects many different environments.

Investigation
Talk about how some things are biodegradable and others are not. Thinking about what will happen to things that we use in our everyday life is a reminder that we all leave environmental footprints and have a responsibility to reduce the impact that we have on the planet.

Lesson 2: 'GREEN LIVING'

How can we reduce pollution?
This lesson considers some of the ways pollution can be addressed. It also introduces the notion of renewable energy. It is important that children are not left feeling helpless. Learning about current issues is the first step towards constructive engagement and action.

Discussion
Begin by asking pupils to work out and explain to each other the methods shown in the images.

Investigation
Pupils could work in groups to devise their policy – the exercise is likely to generate a good deal of discussion.

Climate change
Draw attention to some of the ways that people can reduce their consumption. This has the potential to tackle climate change at its source. The fewer resources we use, the less pollution we will cause. This benefits wildlife and the environment, as pupils will discover when they consider different examples.

Lesson 3: EXPLORING CLEAN ENERGY

Can old power stations make clean energy?
The UK has phased out coal fired power stations in little over a decade. Around 40% of power now comes from renewables, and gas accounts for 20%.

Discussion
Drax power station once consumed huge quantities of coal but now burns wood pellets. As the text explains, this is still problematic. The problem is that all forms of power have some environmental impact. The challenge is to keep this to a minimum.

A local investigation

Investigation
Making a study of pollution in and around the school will alert pupils to local problems at a scale which they can understand. The investigation might stimulate debate about how local communities can contribute to solving problems and improving the quality of the environment.

Mapwork
Demonstrate what the map or plan might look like using the example in the Pupil Book.

Teacher's Guide photocopiable resources
Use pages 45–47 to consolidate key concepts.

Workbook
See pages 32–37 for additional supportive activities.

Unit-by-unit notes

Unit 7: Wales

In this unit, pupils learn:
- what makes Wales different from other countries in the United Kingdom
- about the workings of a coal mine
- how some parts of Wales are changing.

Wales has been politically united with England since 1535 but has retained its tradition and culture, including having the dragon as the national symbol. It has a population of just over three million people, the majority of whom live in Gwent, and Mid, South and West Glamorgan. Around one person in six can speak or understand Welsh. In the past, most jobs in Wales used to be in mining and heavy industry such as steel making, but many factories have now closed. Self-employment and tourism are increasingly important.

Lesson 1: MOUNTAINS AND VALLEYS

What is Wales like?
This lesson introduces pupils to Wales and highlights some of the factors which contribute to its distinct identity. There is a balance between physical and human geography themes in both the text and photographs. The map of Wales shows some of key features which children need to recognise for their locational understanding.

Discussion
Support this discussion by giving a list of points to consider, such as mountains, rivers and landscape, weather, where people live, settlements, work and transport.

Mapwork
Children will need to use string or something flexible to measure the distance round the coast. This might lead to other coastline measurements such as the length of the Welsh coastline, which will enhance their understanding of scale and the use of a scale bar.

Investigation
There are considerable differences between North and South Wales in terms of settlement, transport and work.

Lesson 2: THE STORY OF BLAENAVON

How is Wales changing?
Blaenavon was built as a mining town to provide iron and coal in the Industrial Revolution. The way that it flourished and declined is typical of many other places in South Wales. Change is a major theme throughout this unit.

As pupils learn about Blaenavon, they will also be able to see how people have interacted with their surroundings. The story of the town would have been very different had minerals not been discovered there.

Discussion
You may need to support some pupils by guiding them to think about population size, employment, how towns decline, and how tourism might have revived and affected the town. You could also ask about pollution and wildlife.

Mapwork
If an Ordnance Survey map is not available, it would be possible to make comparisons with Google Maps.

Investigation
Extend this activity by encouraging pupils to think about how Blaenavon might evolve in the future, too.

Lesson 3: A VISIT TO BIG PIT

What was it like to be a coal miner?
Big Pit is on the eastern edge of the South Wales coalfield and one of a number of other similar mines in the area. The deep coal mines closed several decades ago; the opencast mines have closed more recently. Big Pit illustrates a way of life that no longer exists and now serves as a monument to the endeavour that made Great Britain a world power in the 19th and early 20th centuries.

Discussion
Give pupils a chance to think about the experience of visiting Big Pit and what visitors learn; then ask them to discuss in pairs what is important about this. Gather ideas and ask pupils to explain their reasons.

Mapwork
A spoke chart can provide a visual representation of local attractions. Pupils might draw the spokes different lengths according to their distance from the school.

Investigation
World Heritage Sites fall into two categories: they are either buildings and places which have been made by people or they are natural features. As they conduct their research, the children should think about why particular sites have been selected for designation. Always follow the latest advice for practising online safety in research activities.

Teacher's Guide photocopiable resources
Use pages 48–50 to consolidate key concepts.

Workbook
See pages 38–43 for additional supportive activities.

Unit-by-unit notes

Unit 8: Greece

In this unit, pupils learn:
- about the environment of Greece
- what Athens is like in the summer
- about the Greek islands.

Greece consists of the land at the southern tip of the Balkan Peninsula and numerous small islands in the Aegean and Ionian Seas. It is a mountainous country with strong agricultural traditions. The hot, dry summers and extensive coastline have made it popular with tourists. However, there are also environmental problems. Athens suffers from air pollution, and the whole country lies in an earthquake zone. Political links between Greece and other countries were strengthened in 1981 when the country became the tenth member of the European Union. Cultural links date back to ancient Greek civilization which is the source for many contemporary ideas in science, art, politics and philosophy.

Lesson 1: INTRODUCING GREECE

What is Greece like?
You could introduce the unit with a class discussion in which the children list all the things they know about Greece. The exercise will serve as a baseline against which to gauge their progress and development.

Mapwork
This exercise requires pupils to use a scale bar and will help to consolidate their understanding of distance.

Discussion
You could lead a class discussion about the differences between Greece and your country, using the headings in the Pupil Book and anything else they might want to know. This will provide a bridge into the Investigation.

Investigation
You might help pupils structure their fact files by providing them with headings. Alternatively, you might extend it as a literacy exercise.

Lesson 2: SUMMER IN ATHENS

What is the summer like in Athens?
There are many children in Athens who live like Dimitra. Many families live in flats, and in the summer nearly everyone sleeps during the afternoon to avoid the intense heat.

Climate change
You might extend this by discussing what pollutes the air in your area and what can be done about it.

Mapwork
As an extension exercise consider why Athens was a good place to build a city. Possible answers are: (1) the hills provided a defensive position above the plains of Attica; (2) Athens is in a good position for reaching the Aegean islands and different parts of the mainland.

Investigation
Comparing timelines will highlight differences between life in Athens and the pattern of life where you live.

Lesson 3: A GREEK ISLAND

What is it like to visit Amorgos?
Amorgos is typical of many Greek islands. It has a rugged landscape and small ports around the coast. There is also a monastery and some old windmills. Ecologically, the island has considerable conservation value. There are also archaeological remains which date back some 5000 years.

Discussion
Pupils might work in small groups to get a feeling for what it would be like to visit Amorgos in the summer months. They should gather detailed information from Pupil Book pages 48–49 and evaluate what it might contribute to a visit.

Mapwork
Children will have the chance to reflect on the distinct characteristics of Amorgos as they make their own maps.

Investigation
Thinking about what Dimitra might say about Athens and Amorgos is an exercise which requires children to combine their geographical understanding with a sense of empathy. This makes it an effective way of exercising critical thinking.

Teacher's Guide photocopiable resources
Use pages 51–53 to consolidate key concepts.

Workbook
See pages 44–49 for additional supportive activities.

Unit-by-unit notes

Unit 9: North America

In this unit, pupils learn:
- about the landscape of Central and North America
- about everyday life in Jamaica
- about how people can tell you about a place.

North America has a number of distinct geographical regions. One of them, the Caribbean, consists of hundreds of tropical islands, the largest of which are Cuba and Hispaniola. Many of the islands are of volcanic origin and rise to considerable heights. Others are much flatter coral reefs or cays (small, sandy islands above coral reefs). The Caribbean has a unique mixture of cultures and population. Many people are descended from Africans who were enslaved to work on sugar and cotton plantations. The Caribbean is also a biodiversity hotspot with a rich variety of plants and creatures including extensive coral reefs. However, deforestation and pollution are considerable issues.

Lesson 1: INTRODUCING THE CARIBBEAN

What is the Caribbean like?
Discuss what children already know about the Caribbean as a baseline for learning. Use the photographs and text on Pupil Book pages 50–51 to expand these ideas and set the scene for a more in-depth study.

Discussion
You might ask pupils to work in pairs to carefully look at the maps, graphs and pictures and text as a way into this discussion. When you gather responses, develop and extend their understanding by providing information where needed.

Mapwork
Ensure the children understand that the Tropics are those parts of the world where the sun is directly overhead at least once a year. The atlas research will help them to see links between different parts of the world which have similar climates.

Investigation
Severe tropical storms affect all areas of the Tropics. They are called hurricanes in the Caribbean, cyclones in the Indian Ocean and typhoons in the Pacific. Always follow the latest advice for practising online safety in research activities.

Lesson 2: FINDING OUT ABOUT JAMAICA

What is Jamaica like?
This lesson and Lesson 3 focus on Jamaica. This country is geographically small enough for young children to make sense of its physical and human characteristics.

Mapwork
The mapwork exercise asks pupils to find out about other Caribbean islands. If groups of children each selected a different island, you could combine their results in a class display.

Lesson 3: LIVING IN JAMAICA

How is Jamaica changing?
Ingrid Morrison provides a very personal account of her life in Jamaica. Inevitably her experience is limited and individual. Ensure children understand that personal accounts like this provide just one example of life in a country. However, personal portraits of this kind have the advantage of being both detailed and authentic. You may be able to find someone such as parent or guardian who can also talk about their life in Jamaica, which will enable the lesson to develop into a more substantial study to further enrich the lesson.

Investigation
Jamaica has strong links with the UK. As well as thinking geographically, pupils might want to explore other dimensions such as history, music, art and literature.

Sunday in Jamaica

Discussion
You may wish to broaden this discussion to compare to pupils' own Sundays, and ask them to explain which they might like more and why.

Jamaica today

Investigation
Pupils might use the facts they have assembled to create a quiz about Jamaica for children in other classes.

Teacher's Guide photocopiable resources
Use pages 54–56 to consolidate key concepts.

Workbook
See pages 50–55 for additional supportive activities.

Unit-by-unit notes

Unit 10: Africa

> **In this unit, pupils learn:**
> - about the cities and countryside of Africa
> - how links with another school can help provide information
> - about some of the issues in Kenya today.

In terms of geology, much of Africa consists of old, hard rocks which have been covered by sand and other sediments. The continent can be divided into two parts. In the north and west, there is a vast low-lying plateau which includes the Sahara Desert. To the south and east, there is a much higher plateau which is the source of most of the main rivers. Africa has abundant natural resources and a small population in relation to its size.

Lesson 1: INTRODUCING AFRICA

What is Africa like?
Discuss what pupils already know about Africa as a baseline of their learning. Ensure children understand that Africa is a continent. Stories, music, art and festivals are all good starting points for learning more about a geographical area. Learning about physical geography can amplify such studies and lead to a more in-depth understanding of different places.

Discussion
You could extend this discussion by asking pupils to find photographs of the different types of African landscape and of goods from different African countries to show to the class.

Mapwork
This exercise will help pupils learn about African countries through a focus on vegetation and biomes.

Investigation
Children might work in groups and pool their ideas as they discuss the questions for the quiz, to develop group work and communication skills.

Lesson 2: LEARNING ABOUT KENYA

Finding out about Kenya

Discussion
You could support the first two points of discussion with individual or group atlas work, then reporting back in a class discussion. Encourage pupils to look closely at the map and its features on Pupil Book page 58. For the final point of the discussion, pupils could use the headings in the fact file on page 59 to prompt comparisons.
Encourage pupils to explain how their lives are similar to or different from Miriam's. This activity will form a bridge into the writing task in the Investigation.

Making links with a school in Kenya

Investigation
The similarities between lives in very different parts of the world is one of the conclusions which might emerge from this activity. Remind children that personal accounts like Miriam's only provide one example of life in a certain place. In this instance, Miriam's account gives the children information about one young child's life in one small village in Kenya.

A parcel from Kenya

Discussion and Investigation page 60
The map of Kamosong and the items on the display table provide first-hand information about life in this particular village in Kenya. Artefacts of all kinds can be an especially valuable teaching resource because they can be handled and used as evidence. You can build up your own collection using items from local shops and supermarkets. Pupils can be invited to contribute items to any display. When set alongside photographs and commercially-produced study packs, there should no shortage of resources for pupils to interrogate.

Lesson 3: LIVING IN KENYA

How is Kenya changing?
Like most countries all over the world, Kenya is changing fast. The growth of cities and pressure on the environment are universal themes. The way that people are responding to these challenges is part of geography. Presenting pupils with up-to-date information about what is happening will help to keep your teaching meaningful and relevant. This lesson looks to the future and could be supplemented by the latest news reports and online searches. Always follow the latest advice for practising online safety in research activities.

Mapwork
Annotating a map is a useful generic skill which pupils can practise in this exercise.

Investigation
There is a danger that children will get carried away by the design process as they make their advertisements. Ensure that they remember to focus on geographical information.

> **Teacher's Guide photocopiable resources**
> Use pages 57–59 to consolidate key concepts.
>
> **Workbook**
> See pages 56–61 for additional supportive activities.

Photocopiable resource matrix

Unit	Photocopiable resource	Description
Seas and oceans	1 Beneath the surface	Children draw pictures and write sentences to describe plants and animals at different ocean levels.
	2 The ocean environment	Children create a 'picture tower' model about different ocean threats.
	3 Learning about seas	An exercise to identify oceans and seas on a world map.
Wearing away the land	4 Rivers in action	Children colour drawings of erosion, transportation and deposition, write a sentence about each and put them in order.
	5 Preventing flood damage	Children link drawings and descriptions of flood prevention measures.
	6 Finding out about rivers	Children draw some river-surveying equipment and say how it is used.
The seasons	7 Changing seasons	Children draw a tree in the four different seasons and name seasonal clues.
	8 Seasons worldwide	Children compile a rainfall graph for Rome (Mediterranean climate) and compare it to Delhi (monsoon climate).
	9 Seasonal influences	Children consider how school activities are affected by summer and winter weather.
Cities	10 Describing cities	Children complete pictures and speech bubbles about city life.
	11 World cities	An exercise to name and locate cities on a world map.
	12 The story of London	Children complete drawings of different London buildings and sort them into categories.
Jobs	13 Making things	Children create a land-use map of a factory using colours specified in a key.
	14 Different jobs	Children consider the different jobs done in a harbour and decide if they would like to do them or not.
	15 Types of work	An exercise to reinforce the difference between primary, secondary and tertiary activity.

Photocopiable resource matrix

Aim	Teaching points
To show how the ocean changes at different levels.	Discuss how the light, temperature and pressure change with the depth of water.
To illustrate how the oceans are important to us in a variety of ways.	Provide glue and scissors for this activity.
To reinforce locational knowledge about oceans and seas.	This activity could be used to support the mapwork exercise on Pupil Book page 6.
To show that erosion, transportation and deposition are part of a sequence.	Talk about each of these different processes before pupils start this activity.
To consider the effectiveness of different strategies for preventing floods.	Ensure that pupils understand the specialist terms like 'levees' and 'cut-offs' before they start this activity.
To illustrate different types of river-surveying equipment.	Check that children write precise descriptions under their pictures.
To emphasise the different characteristics of each season.	Extend the work by asking the children which season they prefer. You could also read seasonal poems.
To contrast monsoon and Mediterranean climates using climate data.	Check that pupils understand that the graph shows monthly figures that they need to represent in the columns.
To illustrate the effect of the weather on human activity.	Discuss some examples the children might use before they begin the activity.
To show that people have different needs and priorities.	Children will have to infer whether each person likes or dislikes the city from what they say. Some children may need support through talk and prompting questions.
To develop locational knowledge about cities around the world.	The children could add other cities to the map using an atlas to extend the activity.
To consider the different functions and facilities in a major city.	Ensure that pupils realise that each building may perform more than one function.
To introduce pupils to land-use maps using a large-scale example.	Talk about the way that different activities are grouped together in the key before pupils start this activity.
To illustrate the range of work opportunities in a single work environment.	You could extend this activity by considering the range of jobs (both direct and indirect) needed to run your school.
To show that there are major categories of work and that not all work results in a tangible product.	Children could use their own examples of different jobs as well as the examples in the Pupil Book.

Photocopiable resource matrix

Unit	Photocopiable resource	Description
Pollution	16 Damaging the environment	Children play a 'snakes and ladders' game based on environmental themes.
	17 'Green living'	Children draw pictures and write descriptions of different ways of protecting the environment.
	18 Exploring clean energy	A survey sheet that children can use for studying local pollution problems.
Wales	19 Mountains and valleys	Children locate key rivers, mountains and settlements on a map of Wales.
	20 The story of Blaenavon	Children colour and analyse a cross-section diagram of a coal mine.
	21 A visit to Big Pit	Children make a tourist leaflet for Big Pit.
Greece	22 Introducing Greece	Children add colour, titles and descriptions to four pictures showing different aspects of Greece.
	23 Summer in Athens	Children conduct a survey of 12 other pupils to see whether they would prefer to live in the town or country.
	24 A Greek island	Children complete a simple picture diary of Dimitra's visit to Amorgos.
North America	25 Introducing the Caribbean	Children identify and label some of the islands, countries and seas and oceans in the Caribbean.
	26 Finding out about Jamaica	Children label a map of Jamaica and draw pictures and write sentences to show two different scenes.
	27 Living in Jamaica	Children draw pictures to show change in Jamaica, based on the information in the Pupil Book.
Africa	28 Introducing Africa	Children add labels to a blank outline map of Africa.
	29 Learning about Kenya	Children compile an information page with facts and figures about Kenya.
	30 Living in Kenya	Children devise a 'kite diagram' about life in Kenya.

Photocopiable resource matrix

Aim	Teaching points
To introduce pollution problems and solutions in an age-appropriate manner.	If children land on the bottom of a ladder, they go up it. If they land on the tail of a snake, they move back down it. Ensure that pupils understand how the snakes and ladders are linked to real-world issues.
To support children to realise that they can help to protect the environment with their own actions.	Some pupils may need help when they explain how each activity protects the environment.
To promote local fieldwork and investigation.	Find safe places where children can complete the survey or set it for homework to complete with their parents/guardians.
To consolidate points of reference on a map of Wales.	Check that pupils do not obscure the labels and symbols when they colour the map.
To show how coal is often buried deep beneath other rock strata.	Talk with pupils about how the mine works either before or during this activity.
To extract and summarise information from a text passage (in the Pupil Book).	Provide scissors and crayons/coloured pencils/pens.
To reinforce understanding of different geographical themes.	Get the pupils to arrange their drawings around a map of Greece to extend the activity.
To explore attitudes to urban and rural life.	Introduce this activity with a class discussion so that pupils give considered responses when they complete the survey.
To create a visual portrait of life in a Greek island.	Pupils should refer to the text, photographs, maps and captions in the Pupil Book.
To introduce pupils to the geographical setting and context of the Caribbean.	Shading around the islands and coasts in blue will help to make the maps both clearer and more effective.
To familiarise children with different aspects of Jamaica.	Pupils could refer to any of the photographs in this unit in the Pupil Book when they draw their pictures.
To illustrate some of the advantages and disadvantages of modern developments.	Discuss with the children who benefits from change and whether everyone benefits equally.
To introduce pupils to Africa through key physical and human features.	Remind pupils of the names of some of the countries of the continent of Africa.
To build a portrait of Kenya using different types of information.	Check that the children understand the weather chart and the meaning of the different headings before they begin.
To highlight contrasting and complementary aspects of recent changes.	You might get children to work in groups to discuss what they think about recent changes.

1 Beneath the surface

Name ....

1. Draw the plants and animals which live at different levels in the ocean in the empty boxes.
2. Write a sentence about each drawing.

Near the surface _____ _____ _____	
Between 150 and 300 metres _____ _____ _____	
Below 500 metres Below 500 metres there are unusual fish like the Northern wolffish.	

2. The ocean environment

Name ....

Make a 'picture tower' about the threats to the ocean environment.

1. Draw a picture in each panel below.
2. Cut round the edge, fold along the dotted lines and glue the box together.

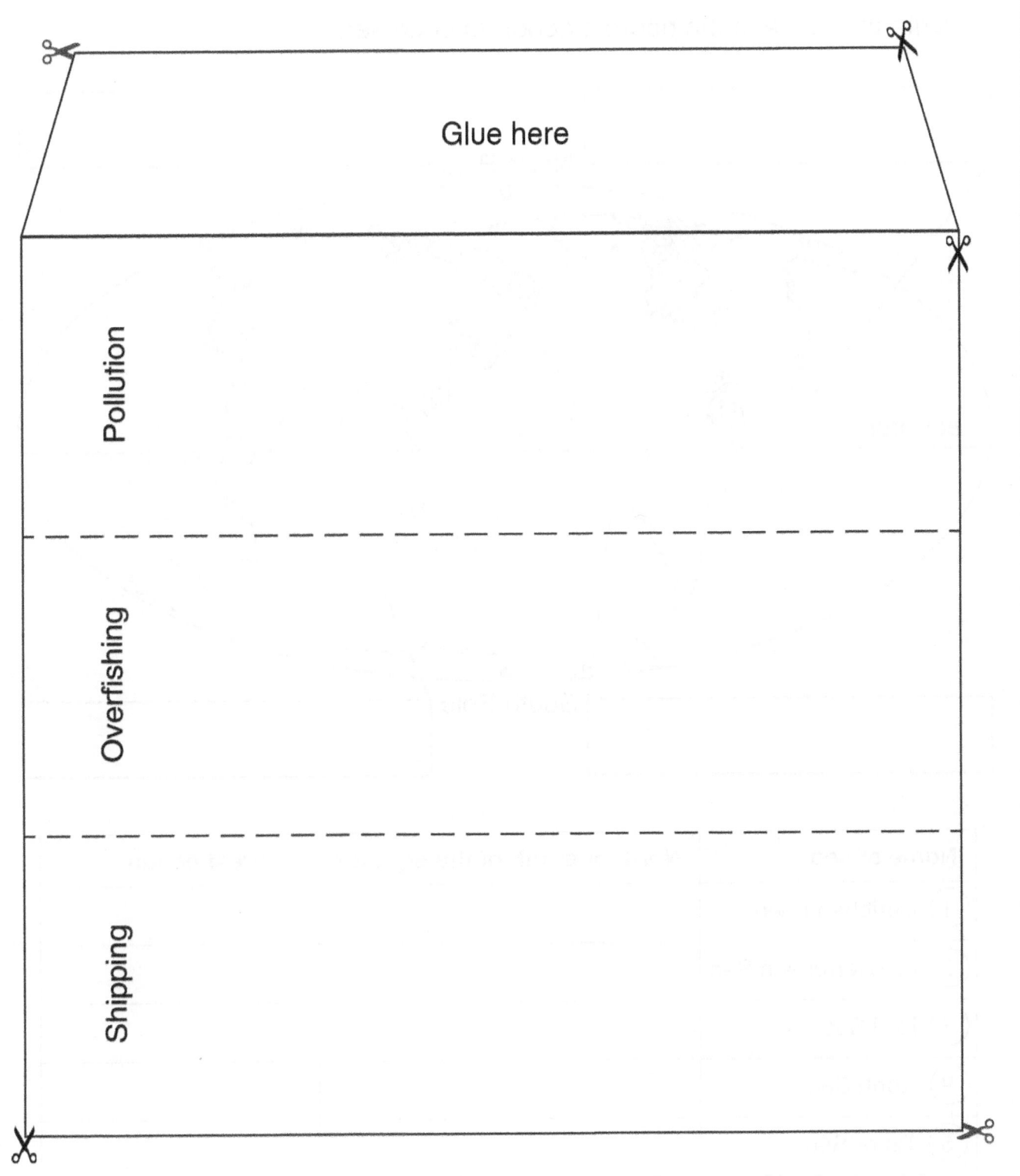

3. Learning about seas

Name ....

1. Write the names of the oceans in the boxes around the map.
2. Find the seas listed in the table on the map.
3. Decide if each sea is north or south of the equator.
4. Write the name of the nearest ocean to each sea

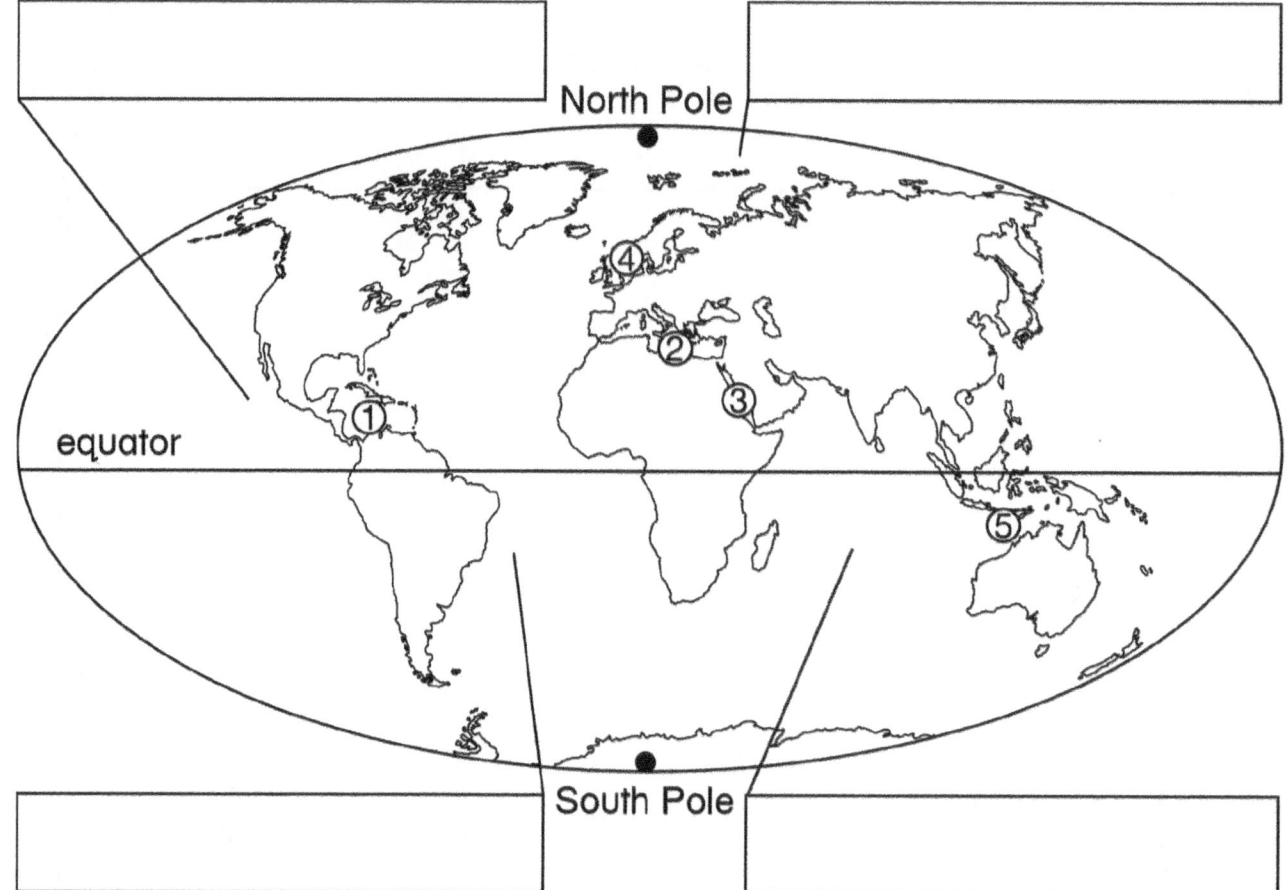

Name of sea	North or south of the equator	Nearest ocean
① Caribbean Sea		
② Mediterranean Sea		
③ Red Sea		
④ North Sea		
⑤ Timor Sea		

4. Rivers in action

Name ....

1. Colour the three pictures.
2. Write a sentence to describe what is happening in each picture.
3. Draw arrows to link them together in the correct order.

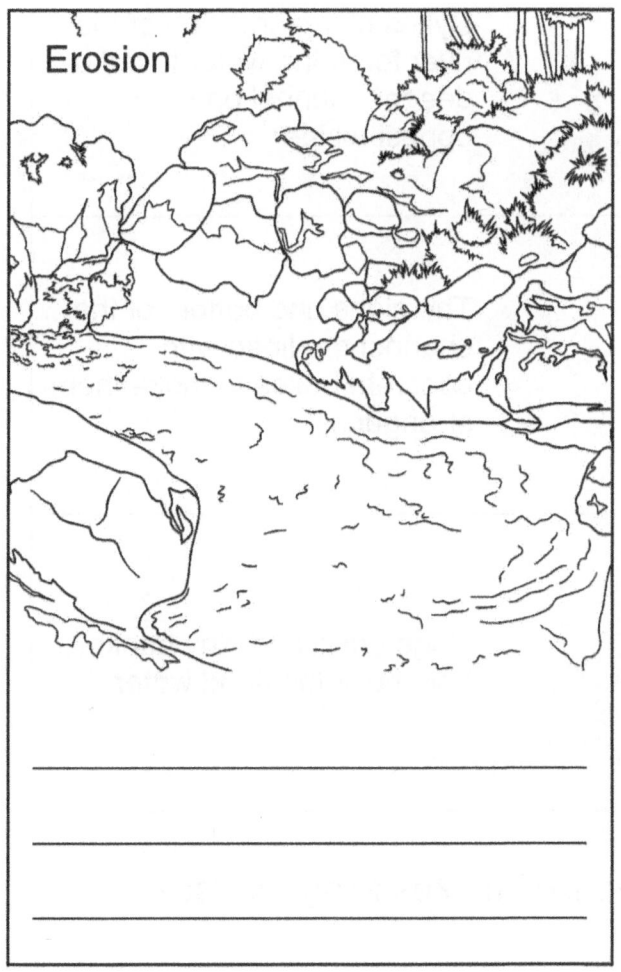

Erosion

_____
_____
_____

Transportation

_____
_____
_____

Deposition

_____
_____
_____

5 Preventing flood damage   Name ....

1. Colour the pictures.

2. Draw lines from each picture to the correct name and description.

	Dykes	New channels cut off some of the meanders so the water can flow faster.
	Levees	Dykes along one side of the river force the water to cut a deeper channel on the opposite side.
	Cut-offs	The sides and bottom of the channel are lined with concrete boxes to make them stronger.
	Boxes	Huge earth and clay banks hold back the flood water.

3. Why do some people think flood prevention makes things worse?

_____

_____

_____

6 Finding out about rivers

Name ....

1. Draw the different pieces of equipment in each box.
2. Write a sentence to explain what each one is used for.

Tape measure	Binoculars	Ranging poles
Coloured corks	Plastic bottle	
Magnetic compass	Information book	

7 Changing seasons

Name ....

1. Draw a tree in spring, summer, autumn and winter.
2. Colour the drawings to show the season.
3. Write one clue which tells you the season shown in each picture.

Spring	**Summer**
Clue: _____	Clue: _____
Autumn	**Winter**
Clue: _____	Clue: _____

8 Seasons worldwide

Name ....

1. Look at the table below. This shows average rainfall in Delhi.
2. Colour the rainfall graph for Delhi.
3. Make and colour your own rainfall graph for Rome.
4. What differences do you notice?

Rainfall in Delhi (monsoon climate)

Month	J	F	M	A	M	J	J	A	S	O	N	D
Rainfall (mm)	33	26	25	14	25	51	233	203	122	18	7	6

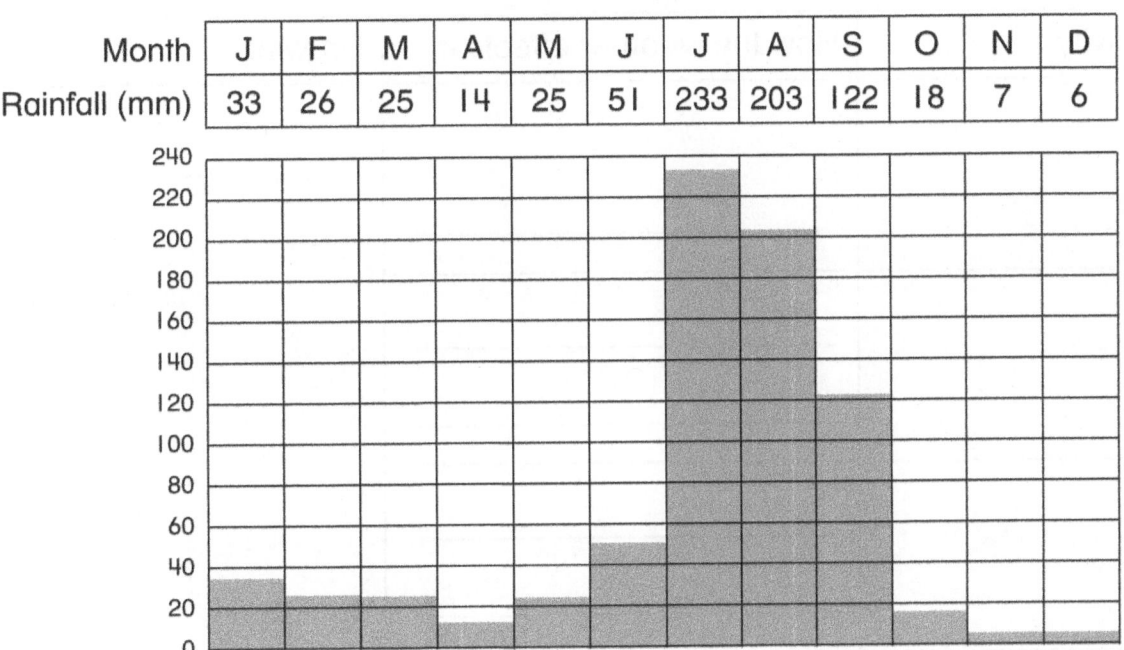

Rainfall in Rome (Mediterranean climate)

Month	J	F	M	A	M	J	J	A	S	O	N	D
Rainfall (mm)	61	52	39	42	49	28	18	27	70	90	95	70

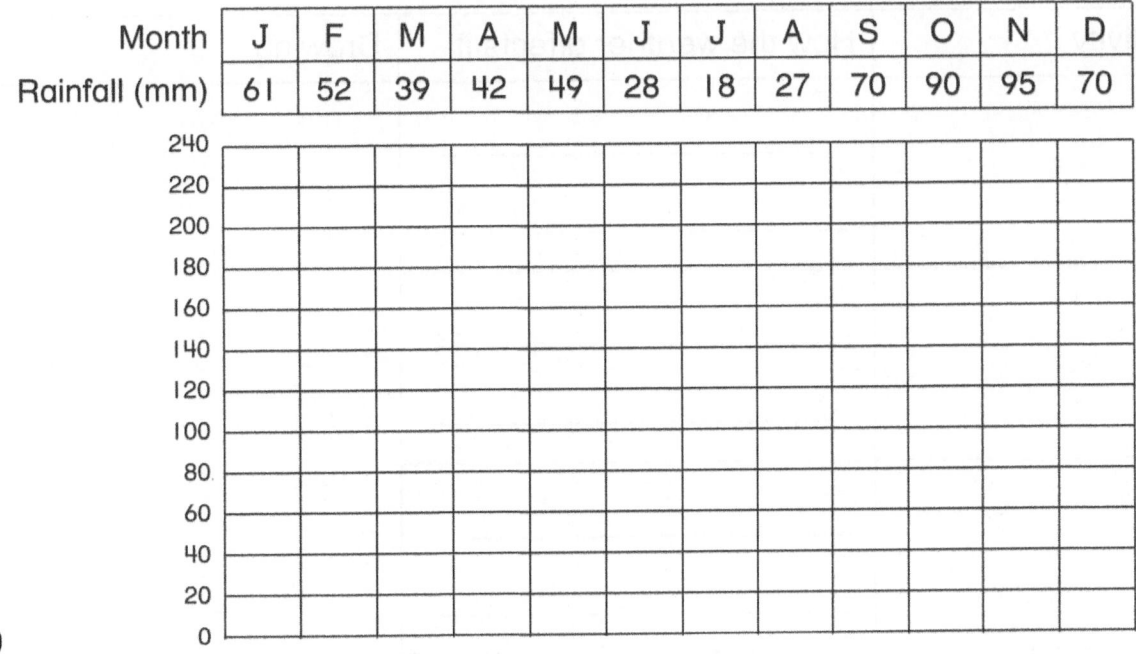

9 Seasonal influences

Name ....

1. Write three things you do at school:
   (a) in summer (b) in winter.
2. Write a sentence which says how each activity is affected by the weather.
3. Draw your favourite summer and winter activity.

Summer

Activity	How the weather affects it	Drawing

Winter

Activity	How the weather affects it	Drawing

10 Describing cities

Name ....

1. Complete the drawings of the people and the words they say.
2. Tick the box to show if the person likes or dislikes city life.

11 World cities

Name ....

1. Write the names of the cities which are marked on the map.
2. Colour the land and the sea.

Name of city	Name of city
1	7
2	8
3	9
4	10
5	11
6	12

12 The story of London

Name ....

1. Draw and write the names of the London landmarks in the empty boxes.
2. Draw lines to link each building to one or more of the circles.
3. Which place do you think is most important?

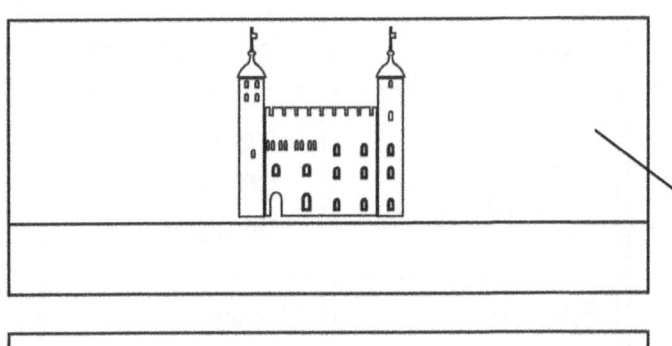

Old Roman Wall

London Eye

Big Ben and Houses of Parliament

○ Tourism

○ History

○ Government

○ Leisure

○ Art

13 Making things

Name ....

1. Colour the empty boxes in the key below.
2. Colour the plan of the factory using the colours from the key.

Key

Work areas	Transport	Energy	Environment
workshop offices storage space	main entrance delivery area car park	boiler chimney	trees grass hedges
Red	Yellow	Brown	Green

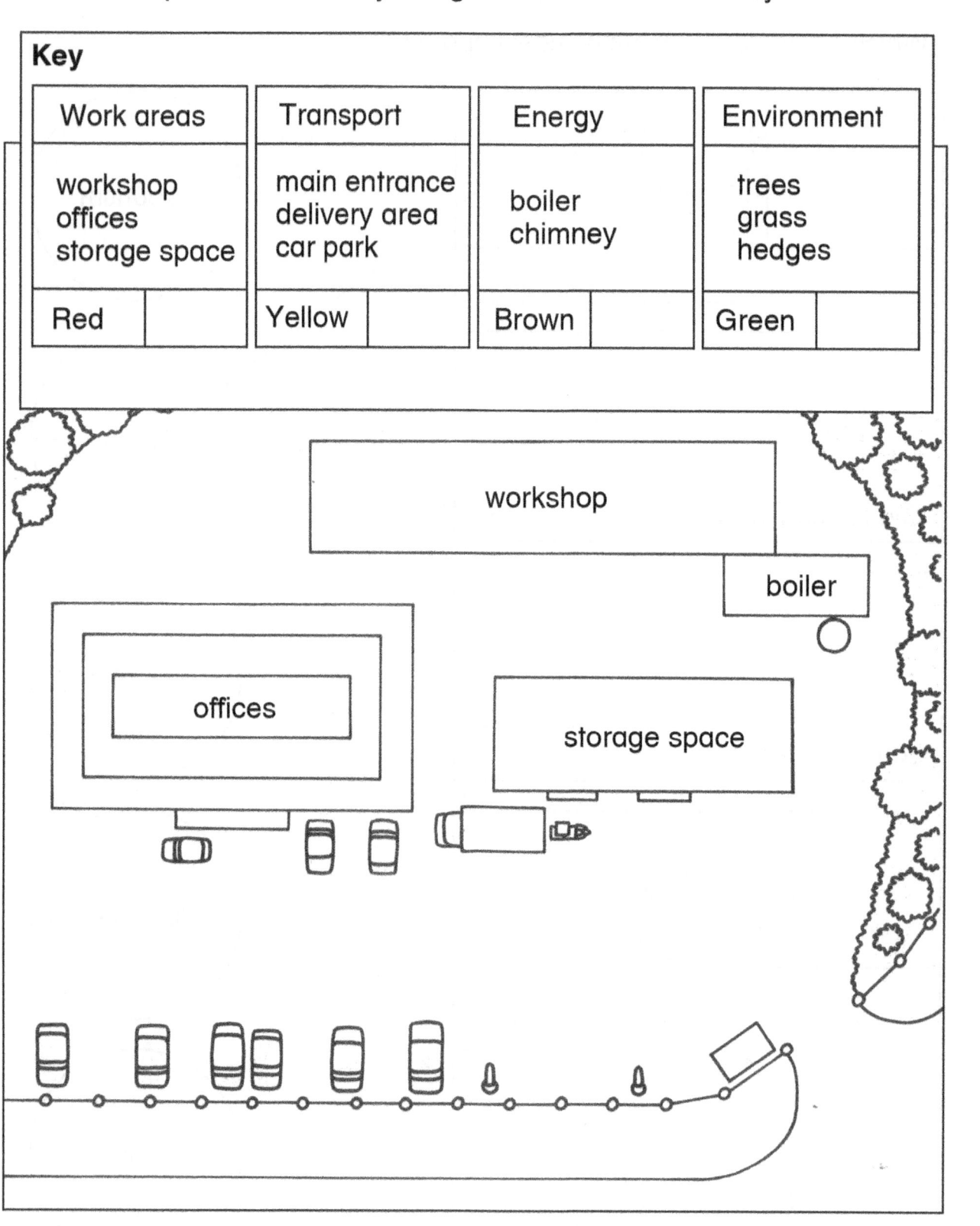

14 Different jobs

Name ....

1. Describe the job done by each person who works in the harbour.
2. Tick one of the boxes to show if you would like to do that job.

Person	What job do they do?	I would like this job	I wouldn't like this job
Joan Lovell			
Carla Gulati			
Bill Shaw			
Steven Bell			
Jack Perez			
Winston Hayes			
Susan Hoff			
Maria Bruni			

15 Types of work

Name ....

1. Draw a picture of one primary, one secondary and one tertiary activity.
2. Write a description of each job.

Primary activity Collecting and harvesting natural resources _____ _____ _____ _____	
Secondary activity Making things _____ _____ _____ _____	
Tertiary activity Providing a service _____ _____ _____ _____	

16 Damaging the environment   Name ....

Play this game with a friend. You will need a 1–6 spinner and counters.

FINISH	55	Dams drown farmland in China	53	52	Deserts spreading	50
43	44	45	Sewage pollutes North Sea	47	48	49
42	41	40	39	38	37	Area of rainforest saved
29	30	31	32	33	34	Global warming
Panda cubs born	27	26	25	24	23	22
15	16	17	Oil spill prevented	19	20	21
14	13	12	11	Lakes poisoned in Canada	9	8
START	More people use electric cars	3	New agreement to protect Antarctica	5	6	7

Collins Primary Geography
Pupil Book 5: Pollution pp32–33

17 'Green living'   Name ....

1. Draw symbols to show each activity in the empty boxes.
2. Write how each activity protects the environment.

Activity	Picture	How it protects the environment
Recycling aluminium cans		_____ _____ _____
Making compost		_____ _____ _____
Travelling on foot or by bike		_____ _____ _____
Using 'green' goods		_____ _____ _____
Saving energy		_____ _____ _____

18 Exploring clean energy   Name ....

Make a pollution survey in your area.

1. Think about each problem and circle one of the numbers.
2. Add up the total score.
3. Write a few sentences about pollution problems in your area.

Problem	No problem	Some	A lot
Traffic exhaust fumes	0	5	10
Factory fumes	0	5	10
Traffic noise	0	5	10
Aircraft noise	0	5	10
Polluted water	0	5	10
Litter	0	5	10
Overhead wires	0	5	10
Unpleasant smells	0	5	10
Factory noise	0	5	10
Vandalism/graffiti	0	5	10
Total			

0 ──────────────────────────────────▶ 50 or more

No pollution   Some pollution   A lot of pollution

_____

_____

_____

_____

19 Mountains and valleys

Name ....

1. Write these labels on the correct place on the map.

Cardiff	Swansea	Bangor	Newport
River Wye	River Severn	Cambrian Mountains	
Anglesey	Bristol Channel	Yr Wyddfa (Snowdon)	

2. Add drawings of the mountains and colour the map.

20 The story of Blaenavon

Name ....

1. Colour this cut-away picture of a coal mine.

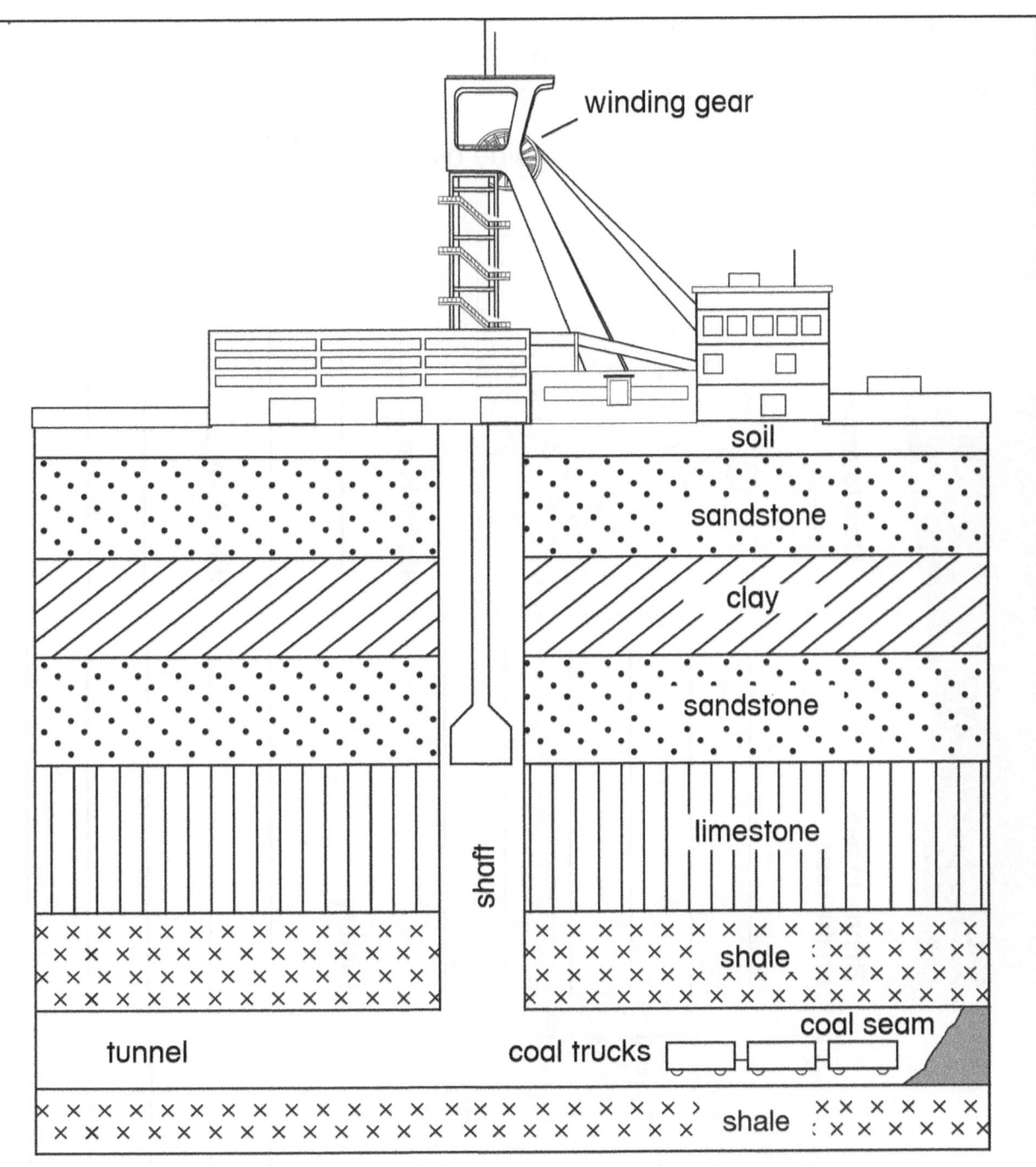

2. What has happened to Big Pit?

_____

_____

_____

21 A visit to Big Pit

Name ....

Make a tourist guide for Big Pit.

1. Colour the pictures and the map.
2. Write in the empty spaces.
3. Cut round the edge and fold your guide along the dotted lines.

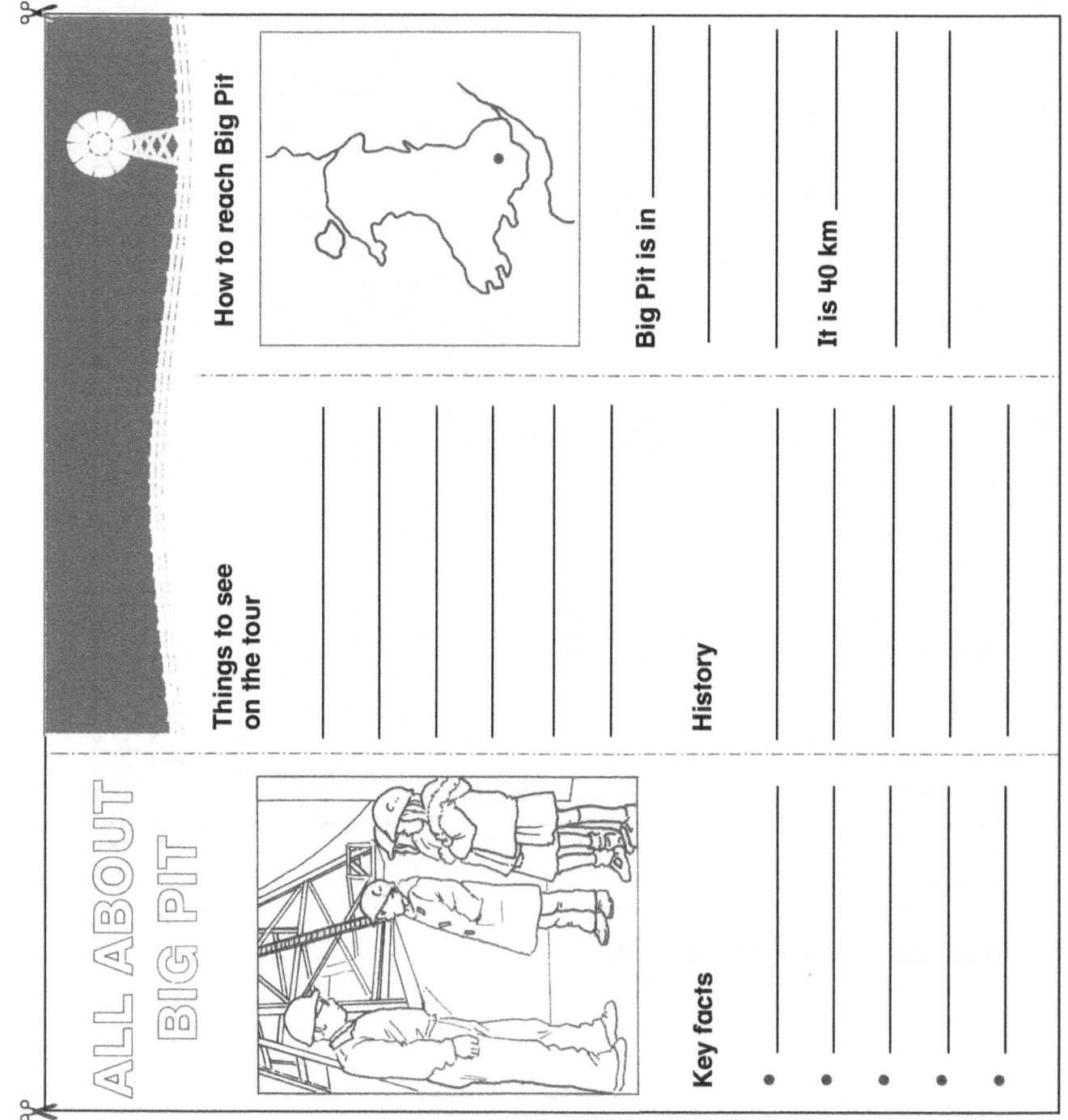

22 Introducing Greece

Name ....

1. Colour the pictures.
2. Write the correct title in the empty boxes.

   | Landscape | Transport | Work | Climate |

3. Write what each picture shows in the space underneath.

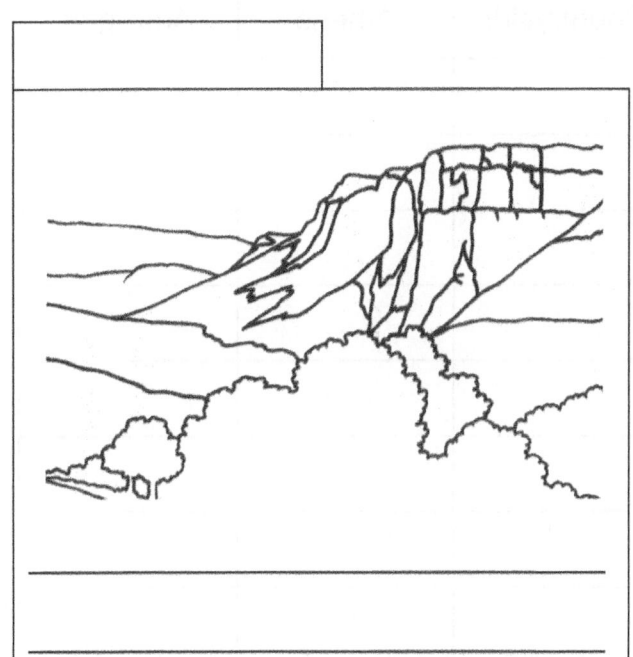

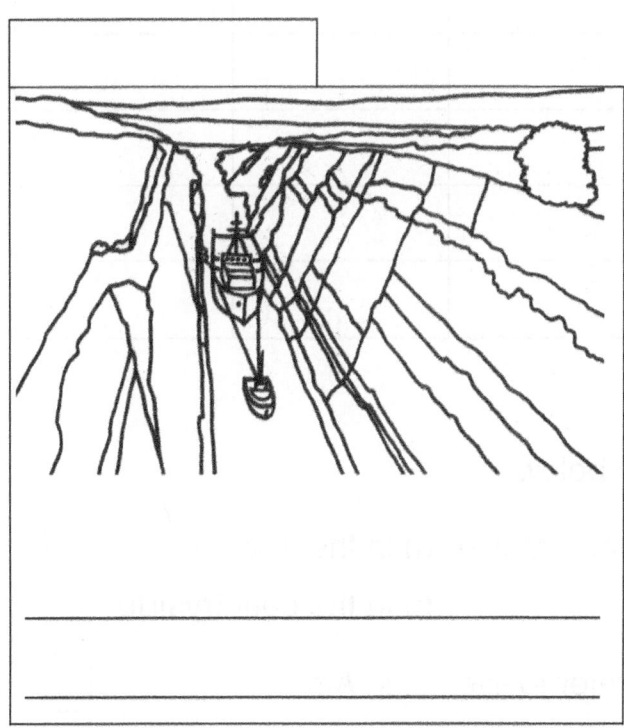

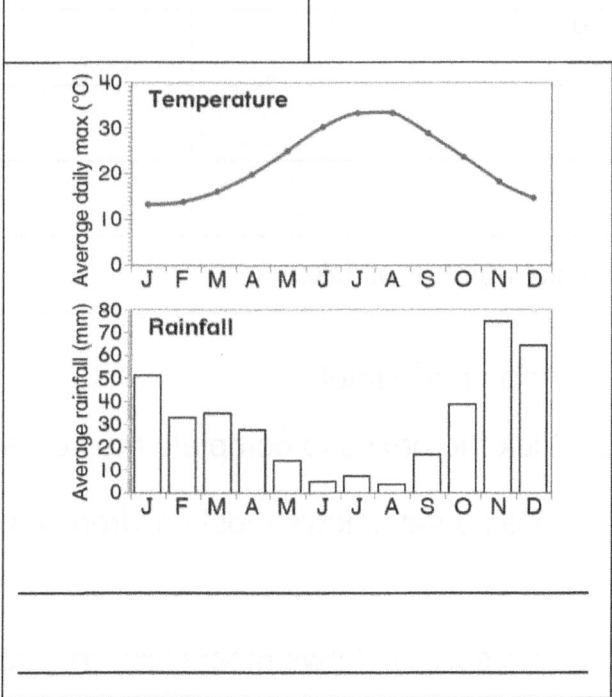

23 Summer in Athens

1. Ask 12 children in your class these questions.

   (a) Would you prefer to live in the town or the countryside?

   (b) Would you prefer to live in Athens or Amorgos?

2. Show their answers by putting a tick or cross in the table.

Name of child	Town	Countryside	Athens	Amorgos
Total number of ticks				

2. Add up the totals.

3. Tick the boxes to complete the sentences below.

   The survey shows most children would prefer to live   a) in the town ☐

   b) in the countryside ☐

   The survey shows most children would prefer to live   a) in Athens ☐

   b) in Amorgos ☐

24 A Greek island   Name ....

1. Colour the drawing of the taverna.
2. Draw and label pictures of three other things Dimitra will remember about her visits to Amorgos.

The taverna

25 Introducing the Caribbean   Name ....

1. Label these islands on the map of the Caribbean:

   Hispaniola   Jamaica   St Lucia   Trinidad

2. Label these countries:

   United States   Venezuela

3. Label these seas and oceans:

   Caribbean Sea   Atlantic Ocean

4. Colour the land and sea.

26 Finding out about Jamaica   Name ....

1. Name the places marked on the map of Jamaica.
2. Draw two pictures showing two scenes from Jamaica.
3. Write a sentence saying what each picture shows.

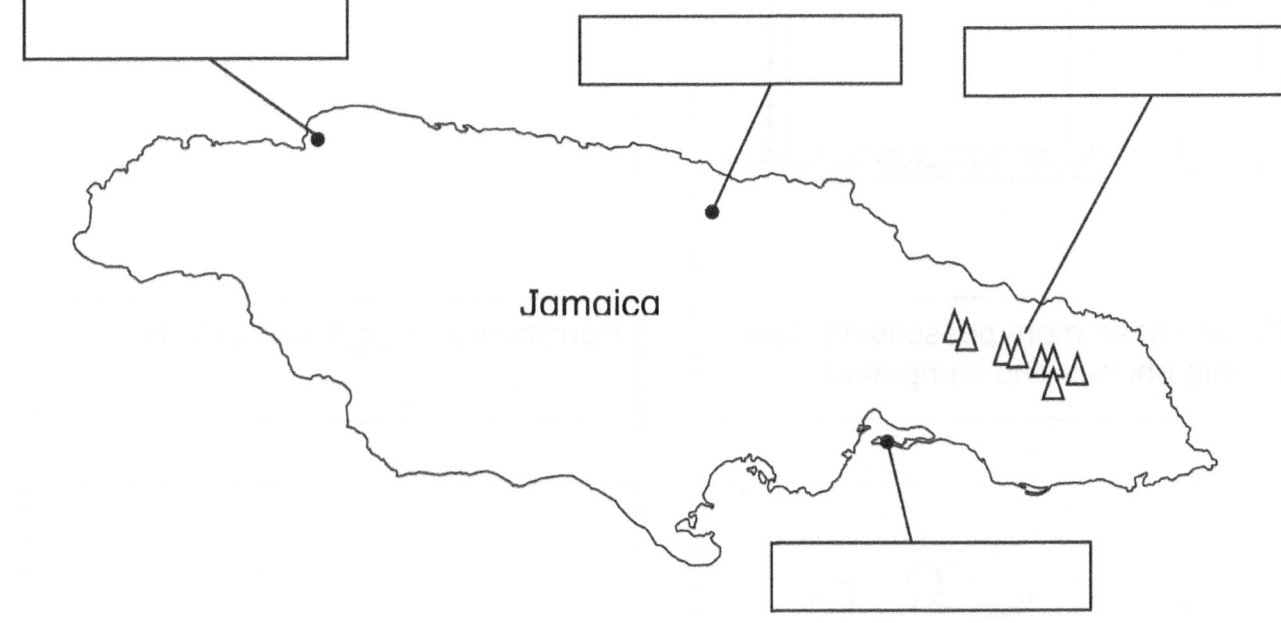

27 Living in Jamaica

Name ....

1. Complete and colour the drawings about how Jamaica is changing.

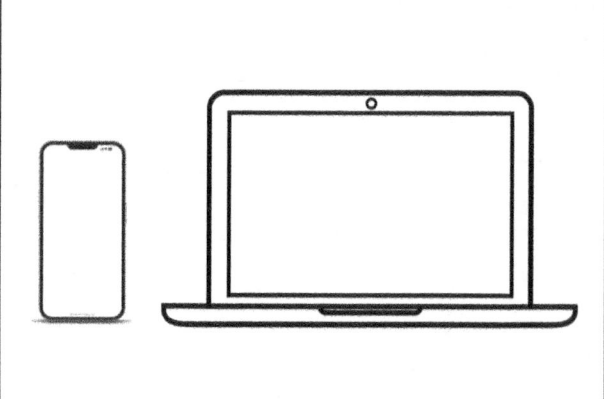

People have more possessions like mobile phones and computers.

Parrots are losing their habitats.

Tourists bring money to Jamaica.

The mangrove forests are being cleared for buildings.

2. Write in your own words how Jamaica is changing.

_____

_____

_____

_____

28 Introducing Africa

Name ....

1. Mark the features and places listed in the table on the map below.

Rivers	Mountains	Deserts	Settlement
Nile	Atlas Mountains	Sahara Desert	Cairo
Congo	Mount Kilimanjaro	Kalahari Desert	Lagos
Zambezi			Cape Town

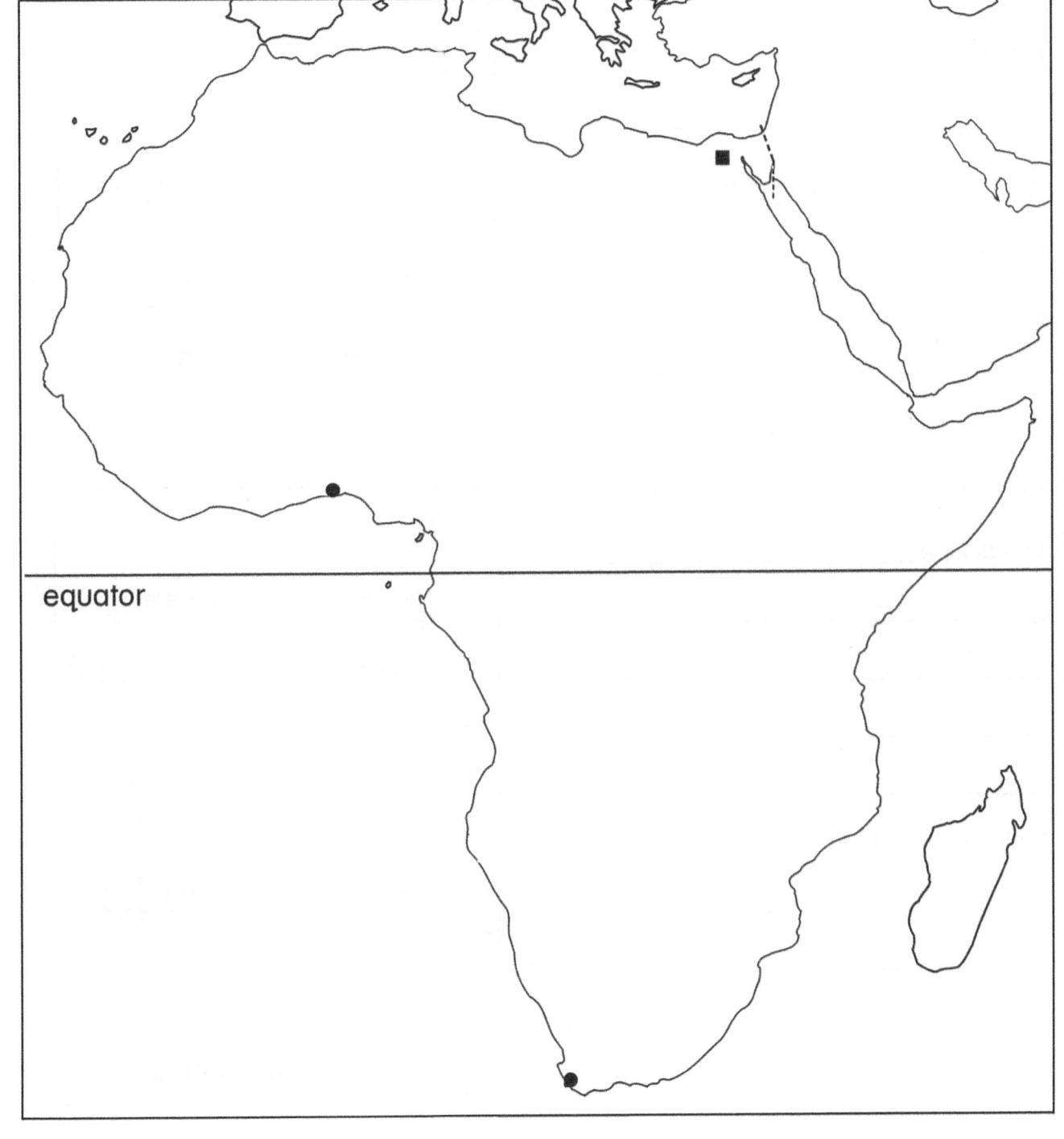

equator

Collins Primary Geography
Pupil Book 5: Africa pp56–57

29 Learning about Kenya

Name ....

Complete the information and colour the map and diagram on Kenya.

Kenya

Population   55 million

Official languages   Swahili and English

Capital city _____

Other cities _____

_____

_____

Rivers

_____

_____

_____

Landscape _____

_____

_____

_____

Main crops _____

_____

_____

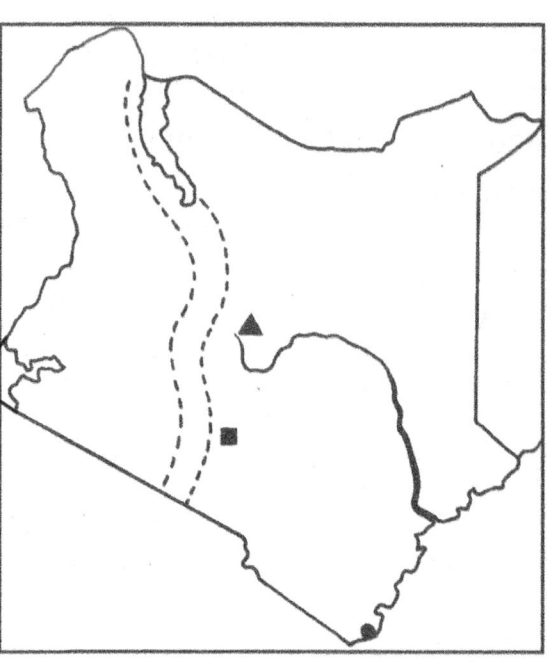

Map of Kenya

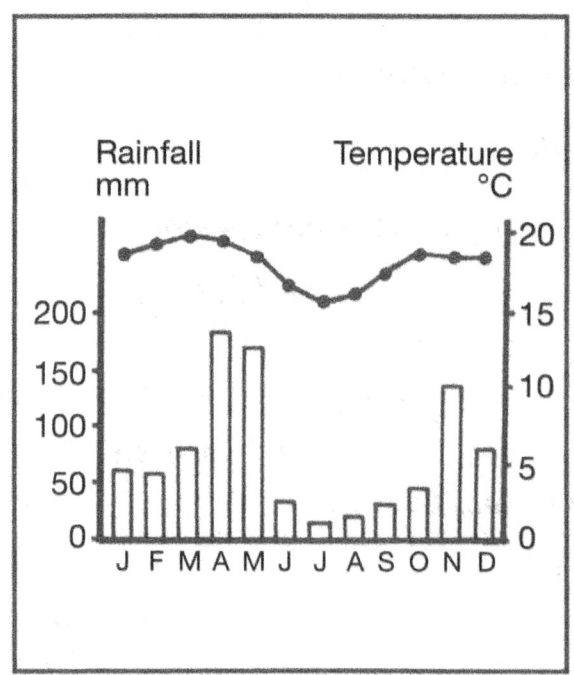

Weather in Nairobi

30 Living in Kenya

1. Write notes and draw pictures about how Kenya is changing on the kite diagram below.

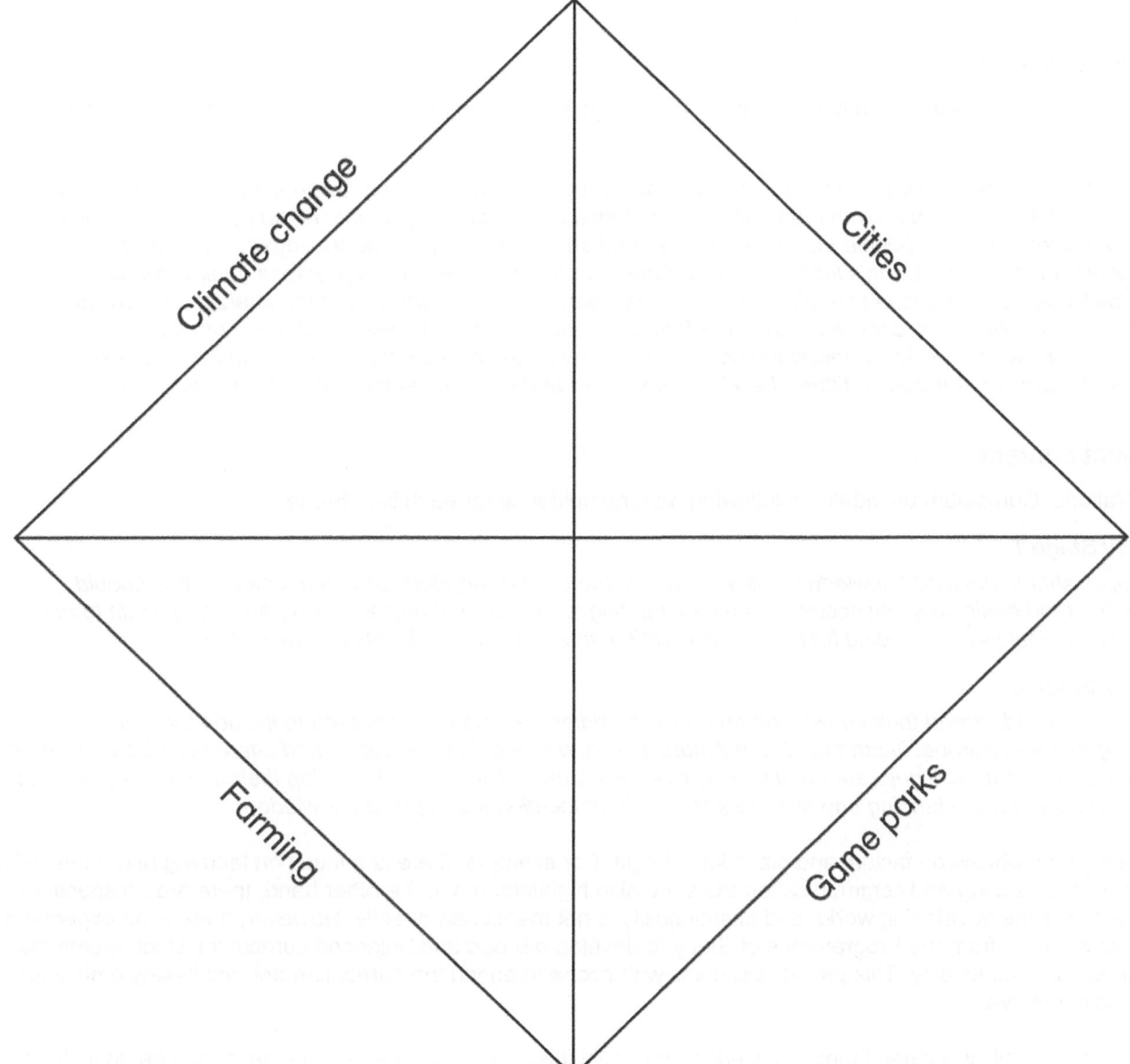

2. Which change do you think is most important?

_____
_____
_____

Geography in the National Curriculum in England

The National Curriculum in England provides a geography framework for schools to follow but leaves teachers considerable scope to select and organise the content according to their individual needs. It should also be noted that the curriculum is only intended to occupy a proportion of the school day and that schools are free to devise their own studies in the time that remains.

Purpose of study

The aim of geographical education is clearly articulated in the opening section of the Programme of Study, which states:

A high-quality geography education should inspire in pupils a curiosity and fascination about the world and its people that will remain with them for the rest of their lives. Teaching should equip pupils with knowledge about diverse places, people, resources and natural and human environments, together with a deep understanding of the Earth's key physical and human processes. As pupils progress, their growing knowledge about the world should help them to deepen their understanding of the interaction between physical and human processes, and of the formation and use of landscapes and environments. Geographical knowledge, understanding and skills provide the frameworks and approaches that explain how the Earth's features at different scales are shaped and interconnected and change over time.

Subject content

The National Curriculum provides the following general guidance for each Key Stage:

Key Stage 1
Pupils should develop knowledge about the world, the United Kingdom and their locality. They should understand basic subject-specific vocabulary relating to human and physical geography and begin to use geographical skills, including first-hand observation, to enhance their locational awareness.

Key Stage 2
Pupils should extend their knowledge and understanding beyond the local area to include the United Kingdom and Europe, North and South America. This will include the location and characteristics of a range of the world's most significant human and physical features. They should develop their use of geographical knowledge, understanding and skills to enhance their locational and place knowledge.

There is an emphasis on factual and place knowledge. For example, there is a focus on learning about the UK and Europe. Map reading and communication skills are also highlighted. On the other hand, there are no specific references to the developing world, and sustainability is not mentioned directly. However, there is an expectation that schools will work from the Programmes of Study to develop a broad and balanced curriculum which meets the needs of learners in their locality. This provides schools with scope to enrich the curriculum and rectify any omissions which they may perceive.

Key Stage 2 pupils will extend their knowledge and apply their skills to areas beyond the local area, to include the UK, Europe, North and South America, Africa and Asia.

Fieldwork is covered throughout the *Collins Primary Geography* series, and consistent opportunities are provided in Investigation and Mapwork activities to 'observe, measure, record, and present the human and physical features in the local area using a range of methods, including sketch maps, plans, graphs and digital technologies', as specified in the National Curriculum.

Key Stage 2 Programme of study

Key Stage 2 Geography National Curriculum	*Collins Primary Geography* coverage
Extend knowledge of UK, Europe and North and South America	Places (all)
Location of world's most significant human and physical features	(all)
Knowledge, understanding and skills to enhance locational and place knowledge	(all)
Locational knowledge	
Locate the world's countries	(all – Mapwork)
Use maps to focus on countries, cities and regions in Europe	Places (all)
Use maps to focus on countries, cities and regions in North America	Book 4, Book 5: Places
Use maps to focus on countries, cities and regions in South America	Book 3, Book 6: Places
Name and locate counties of the UK	Places (all)
Name and locate cities of the UK	Places (all)
Geographical regions of the UK	Places (all)
Topographical features of the UK, such as hills, mountains, coasts and rivers.	Places (all)
Changing land use patterns of the UK	Places (all)
Significance of latitude and longitude	Book 6: Planet Earth
Significance of Equator, Northern and Southern Hemisphere, Tropics of Cancer/Capricorn, Arctic/Antarctic circles, Prime Meridian	Book 6: Places
Time zones	Book 4: Places
Day and night	Book 4: Places
Place knowledge	
Regional study within UK	Places (all)
Regional study in a European country	Places (all)
Regional study in North America	Book 4, Book 5: Places
Regional study in South America	Book 3, Book 6: Places
Human and physical geography	
Climate zones	Weather (all)
Biomes and vegetation belts	Environment (all)
Rivers and mountains	Book 3, Book 6: Planet Earth; Book 4, Book 5: Water; Places (all)
Volcanoes and earthquakes	Book 5, Book 6: Planet Earth; Book 5: Environment; Book 3, Book 6: Places
Water cycle	Book 5: Water
Types of settlement and land use	Settlements (all)
Economic activity including trade links	Work and travel (all)
Distribution of natural resources including energy, food, minerals, water	Book 6: Water; Work and travel (all)
Skills and fieldwork	
Use maps, atlases, globes and digital mapping	(all – Mapwork)
Use eight points of the compass	Book 5: Water
Use four and six-figure grid references	Book 3: Work and travel; Book 4: Settlements
Use symbols and keys (including OS maps)	Book 3, Book 5: Work and travel; Book 3, Book 4: Places
Fieldwork skills	(all – Investigation)

WORLD MAP

WORLD COUNTRIES

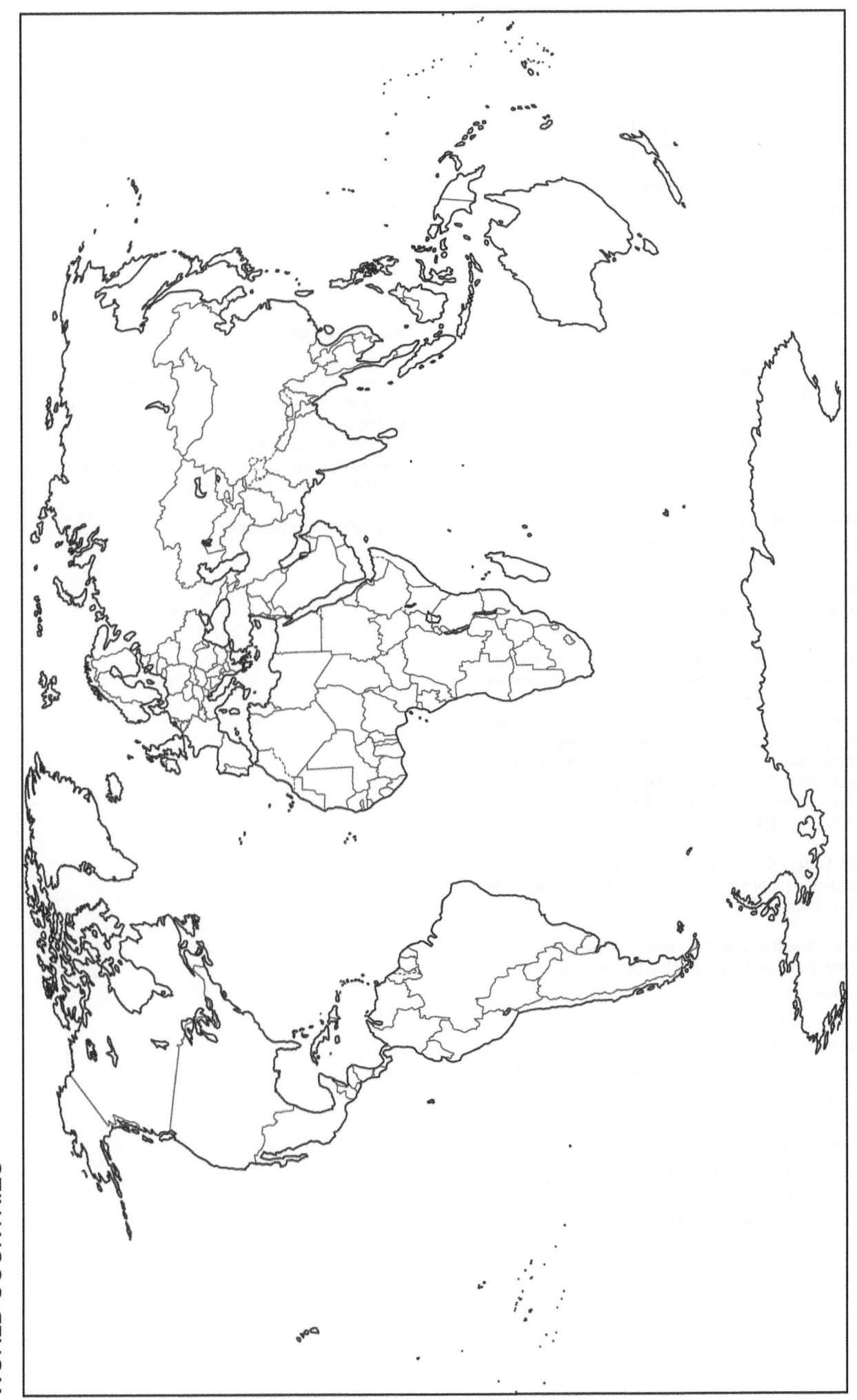

William Collins' dream of knowledge for all began with the publication of his first book in 1819. A self-educated mill worker, he not only enriched millions of lives, but also founded a flourishing publishing house. Today, staying true to this spirit, Collins books are packed with inspiration, innovation and practical expertise. They place you at the centre of a world of possibility and give you exactly what you need to explore it.

Published by Collins
An imprint of HarperCollins*Publishers*
The News Building, 1 London Bridge Street, London, SE1 9GF, UK

HarperCollins*Publishers*
Macken House, 39/40 Mayor Street Upper, Dublin 1, D01 C9W8, Ireland

Browse the complete Collins catalogue at **collins.co.uk**

© HarperCollins*Publishers* Limited 2025
Maps © Collins Bartholomew 2025

10 9 8 7 6 5 4 3 2 1
ISBN 978-0-00-872844-1

All rights reserved. No part of this publication may be reproduced, stored in a retrieval system, or transmitted in any form by any means, electronic, mechanical, photocopying, recording or otherwise, without the prior written permission of the Publisher or a licence permitting restricted copying in the United Kingdom issued by the Copyright Licensing Agency Ltd, 5th Floor, Shackleton House, 4 Battle Bridge Lane, London SE1 2HX.

British Library Cataloguing-in-Publication Data
A catalogue record for this publication is available from the British Library.

Authors: Stephen Scoffham and Colin Bridge
    (with additional original input by Terry Jewson)
Publisher: Laura White
Product manager: Natasha Paul
Development editor: Judith Walters
Proofreader: Hugh Hillyard-Parker
Cover designer and illustrator: Steve Evans
Internal illustrator: Jouve India Private Ltd
Typesetter: Hugh Hillyard-Parker
Production controllers: Alhady Ali and
    Katie Jean-Baptiste
Printed in the UK at Ashford Colour Ltd

This book contains FSC™ certified paper and other controlled sources to ensure responsible forest management.

For more information visit: www.harpercollins.co.uk/green

Acknowledgements

The publishers gratefully acknowledge the permission granted to reproduce the copyright material in this book. Every effort has been made to trace copyright holders and to obtain their permission for the use of copyright material. The publishers will gladly receive any information enabling them to rectify any error or omission at the first opportunity.

All photos: Shutterstock